If You're an Inventor, You Need This Book!
Here are some of the things you will learn:

The Different Types of Inventions	*Page 4*
Three Keys to Protection	*Page 50*
How to File for a Patent	*Page 60*
How to Find U.S. Government Buyers	*Page 113*
How to Target Your Customers for More Sales	*Page 124*
How to Get Press Results *Fast!*	*Page 140*
The Process for Handling a Press Event	*Page 145*
How to Write a News Release	*Page 158*
How to Put Together a Direct-Mail Package	*Page 170*
How to Write a Sales Letter	*Page 171*
How to Use Magazines to Your Advantage	*Page 192*
How to Read Media Rate Cards	*Page 201*
How to Plan a Trade Show	*Page 210*
Trade Show Budgeting	*Page 212*
Five Key Rules in Packaging	*Page 236*
How to Name Your Product	*Page 239*

Also by Reece A. Franklin

101 Ideas for News Releases
Promotion and Sales Ideas for Retail Stores
Inventor's Resource Directory
How to Market Your Home-Based Business

How to Order:

Single copies may be ordered from Prima Publishing, P.O. Box 1260BK, Rocklin, CA 95677; telephone (916) 786-0426. Quantity discounts are also available. On your letterhead, include information concerning the intended use of the books and the number of books you wish to purchase.

HOW TO SELL AND PROMOTE YOUR IDEA, PROJECT, OR INVENTION

An Excellent Marketing Guide
for Both Novice
and Seasoned Inventors

Reece A. Franklin

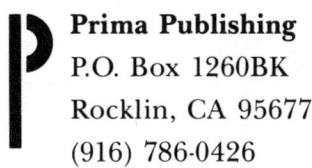

Prima Publishing
P.O. Box 1260BK
Rocklin, CA 95677
(916) 786-0426

© 1993 by Reece A. Franklin

All rights reserved. No part of this book may be reproduced or transmitted in any form or by any means, electronic or mechanical, including photocopying, recording, or by any information storage or retrieval system, without written permission from Prima Publishing, except for the inclusion of quotations in a review.

Production by Jane Brundage, Bookman Productions
Copyediting by Betsy Dilernia
Interior design by Paula Goldstein, Bookman Productions
Cover design by Kirschner-Caroff Design
Typography by Bookends Typesetting
Illustrations by Bookman Productions

Library of Congress Cataloging-in-Publication Data

Franklin, Reece A.
    How to sell and promote your idea, project, or invention / Reece
  A. Franklin
    p. cm.
    Includes index.
    ISBN 1-55958-295-2
    1. Inventions—Marketing.   I. Title.
  T212.F73 1993
  658.8—dc20   92-40481
                                                                                         CIP

---

93 94 95 96 97 RRD 10 9 8 7 6 5 4 3 2 1
Printed in the United States of America

Dedication

To all inventors and entrepreneurs, from whatever country you call home. In your zest to create, you shine above all others. And to all the loved ones of inventors and entrepreneurs: You are the beacons that cast forth that light.

Special Dedication

To Frances Singman (1918–1983). The light still shines, Mom!

A Note to the Reader

This book was written for the purpose of providing information. It is sold with the understanding that the publisher and author are not engaged in rendering legal, accounting, patent, or other professional services. If legal or other expert assistance is required, the services of a competent professional should be sought.

It is not the intent of the author to claim that by reading this book, an inventor may become wealthy from using the information herein. How one uses the information is up to the individual reader.

The publisher and author claim only that the material contained herein is based on currently accepted practices and was learned through the author's research and personal experience. The techniques may not necessarily work for all readers.

Every effort has been made to make this book as complete and accurate as possible. The book should be used only as a general guide, not as the ultimate source of invention development and marketing. This book contains information on invention marketing that is current up to the printing date.

The purpose of this book is to educate and entertain. The author and Prima Publishing will have neither liability nor responsibility to any person or entity with respect to any loss or damage, caused or alleged to be caused directly or indirectly by the information contained in this book.

About the Author

Reece Franklin has had a love affair with inventions and inventors ever since his mother introduced him to the man who invented the Lava Lite.

Reece has been a fund raiser, advertising sales rep, publicist, marketer, and sales manager. He previously owned Singman Publications, a specialty trade publishing and radio communications company in San Diego, California. Prior to that, he was founder of Reece Franklin Public Relations and its subsidiary, Franklin Television, a local cable television production company.

He gives over 100 seminars per year throughout California, including How to Market Your Small Business, How to Get Free Publicity, How to Advertise and Promote Your Small Business, and Designing Effective Small Business Ads.

Reece's new seminar, Inventor's Marketing Workshop, based on this book, is already the talk of the industry.

Currently President of Reece Franklin and Associates in Chino Hills, California, a small business consulting firm specializing in small business start-ups and inventors, he is sought after as a speaker and guest lecturer.

Reece was prompted to write this book by the hundreds of students at his seminars throughout the community college system in Southern California.

Now he reveals to you the secrets of creating, patenting, and marketing your invention in readable, step-by-step chapters. Wait no longer! Open to Chapter 1, and begin your journey!

Contents

Acknowledgments   x

Introduction   xi

PART I In the Beginning   1

1. What's an Inventor?   3
2. The Right Stuff: The Skills You Need   9
3. Getting Started Right: The Basics of Business   20

PART II I've Got an Idea!   37

4. From Idea to Product   39
5. Protecting Your Invention   48

PART III The Money Game   67

6. Financing Your Invention: From Seed Money to Expansion   69
7. Budgeting and Scheduling   81

PART IV Selling Your Invention   91

8. Selling Out for Big Profits: Finding a Buyer   93
9. Distribution: Moving It on Out!   102
10. Selling to the U.S. Government   111

PART V To Market, To Market   121

    11   Marketing: What It's All About   123
    12   How to Get Free Publicity   138
    13   Advertising: Shouting Your Message Out Loud   161
    14   Trade Show Secrets of the Pros   209
    15   Networking: A Little Help from Your Friends   225
    16   Packaging Your Invention   234
          Appendix: Publicity Forms   241

Afterword   253
Bibliography   254
Index   259

Acknowledgments

The hardest part of writing a book comes here, the acknowledgments. It's hard because there are so *many* people to thank, and I may miss someone. I also feel it is inadequate to thank the people who touch my life every day on a mere page of a book. To those of you who are part of my world, no matter what role you play, know that your contributions to this volume are legion, and most humbly appreciated. Without you, this would still be an idea. With you, it's become reality—ready to market!

My gratitude goes to Nadine Johnson, Cheri and David Browner, Laurie and Don Smyers, Nathan Franklin, Dan Poynter, all the Community College and Cal State Extended Education Directors, and *all my students.* Special thanks to Gordon Burgett, for contributions above and beyond the call of duty.

Introduction

This book has been written to fill a need. In marketing circles, one of the key principles is, "Find a need and fill it." I believe there is a need here.

I've met hundreds of inventors and entrepreneurs who have had the "perfect" product, the proverbial "better mousetrap," only to see it lie in boxes on their garage floors. I have also met the inventors of the strangest items, who were doing quite well selling their products both domestically and overseas.

What's the difference? *Marketing!*

Marketing is what makes or breaks a product's life cycle. Without good, solid marketing, nothing gets sold. "It's all in the packaging!" a wise man once said.

My purpose is to make you that wise person—to take you, fellow inventor and entrepreneur, on a path to success. Will you be successful if you read this book? Perhaps—*if* you apply the principles.

Most inventions never get a good return on their investment. The reason is simple. Most inventors do not know how to market their invention.

In contacting inventors' clubs while doing research for this book, I discovered that what is true in my club is true all over the country: When it comes to such things as creating, prototypes, and patents, there are many sources to help the inventor. When it comes to marketing inventions (not selling to corporations, but self-marketing), the number of available books is minuscule. This book solves that problem.

This book is one of the first to focus on the marketing and promotion aspects of your invention or product. In addition to material about legal protection, prototypes, and test-marketing (the necessary starting points), there are full chapters on financing, distribution, selling to the U.S. government, marketing, public relations, advertising, trade shows, and networking—something you won't see in any other inventor's book. In short, this book is a mini marketing course *loaded* with high-power information.

Depending on where you are in your invention's product cycle, you'll find useful information and tips here. And you don't necessarily have to start with Chapter 1.

Finally, I've included information on where to go for help in every area of the product cycle, from starting a business to money sources to inventors' clubs and associations.

So, sit down in your easy chair, kick off your shoes, put on some background music, and let's get started on helping you make money on your invention the old-fashioned way—marketing!

PART I

IN THE BEGINNING

1

What's an Inventor?

Before you get started on your journey to invention marketing, you should answer some basic questions. The discussion that follows will help you understand the types of inventions and inventors there are, and the processes involved in determining if you should invent. Ask yourself:

1. What's an inventor?
2. What is an invention?
3. Why should I invent at all?
4. Where do inventions come from?

WHAT'S AN INVENTOR?

Inventors are unique people. Part psychologist, part scientist, part artist, and hopefully, part businessperson, they are everything good about our country rolled into one.

Inventors, like most people, have their own idiosyncrasies. But inventors understand that these eccentricities help them create. And their creations can lead to the changing of the planet.

By understanding what an inventor is, you will gain some insight into what makes you tick. It is the inventor who dares to challenge the norm, who will create the cure for cancer tomorrow.

Entrepreneurs vs. Inventors

All inventors are entrepreneurs, but not all entrepreneurs are inventors. Inventors need certain entrepreneurial traits in order to succeed. The entrepreneurial inventor (my phrase) is one who starts with a raw idea, turns the concept into a marketable product, and proceeds to promote it through his or her own business—in other words, someone who handles every step of the process. The flowchart in Figure 1.1 outlines the process.

What are the traits of the successful entrepreneurial inventor? Some people believe these would include creativity, a good business sense, an engineering background, financial understanding, along with marketing, accounting, and selling skills. While I agree with this list, I would add a few more characteristics: tenacity, self-discipline, and patience. In addition, there are a dozen mental attitudes you need if you intend to succeed. These are discussed in Chapter 2.

WHAT'S AN INVENTION?

According to the Patent and Trademark Office (PTO) in Washington, DC, patents are granted to inventions that prove "new, useful, practical, workable, and are past the idea stage." If your invention doesn't meet these criteria, I suggest it might not be a *sellable, marketable product* that can make you money!

There are several categories of inventions. They are combinations, labor-saving devices, practical solutions to old problems, or giving new twists to old items. (For further discussion, see Chapter 4.)

Combinations

A combination type of invention results from taking two or more objects and putting them together to create a new product or object. For example,

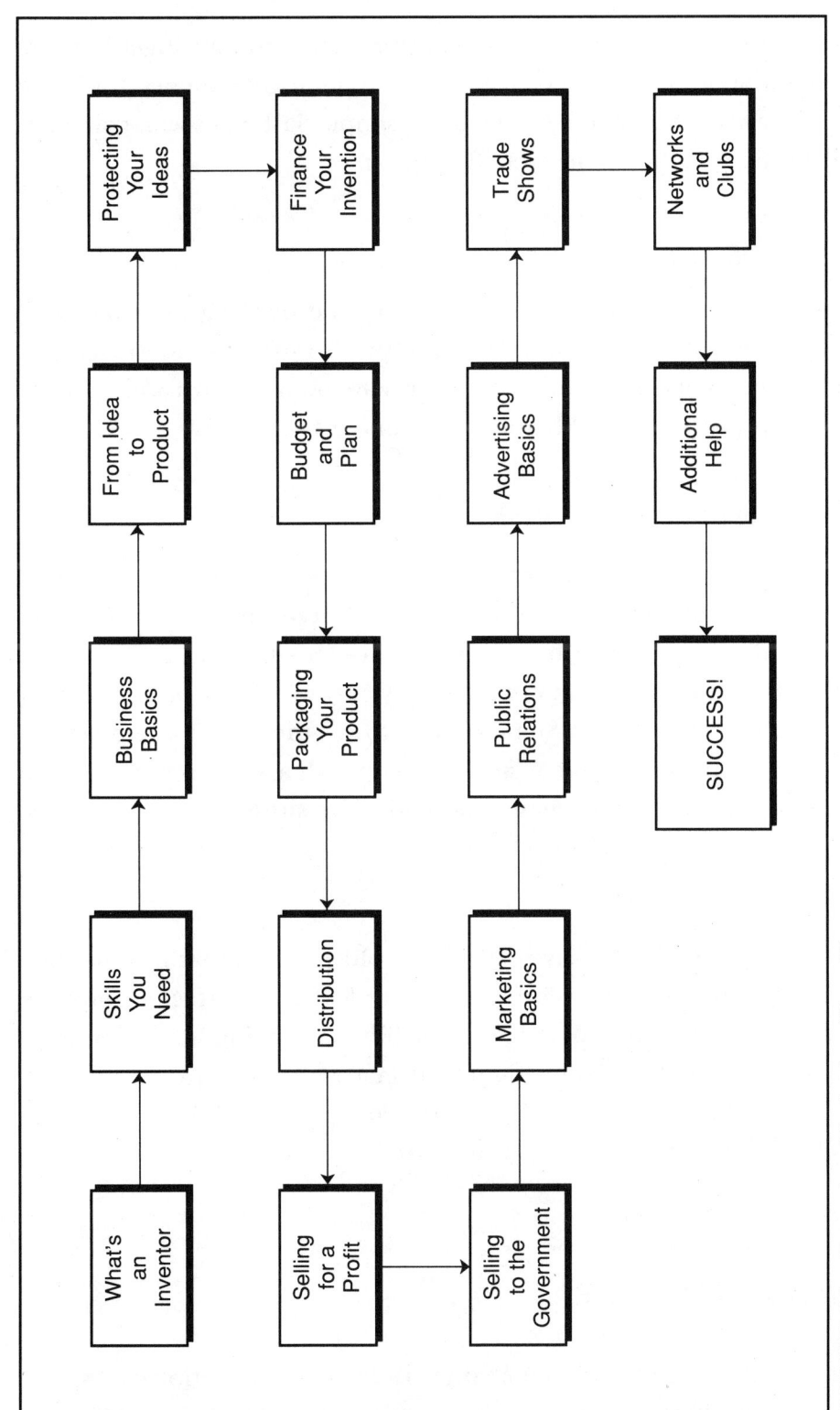

FIGURE 1.1 Invention Marketing Flow Chart

several years ago, after the movie *Star Wars* came out, an ingenious entrepreneurial inventor named Gary Brennan put together a new toy based on the Laser Sword from the movie. Using simple flashlights with golf tubes glued to the end, he made over $1 million.

Labor-Saving Devices

Labor-saving devices generally come under the heading of appliances, although any device that saves time and effort is worth considering. Some common ones that come to mind are microwave ovens, electric can openers, and food processors and blenders, to name a few.

Practical Solutions to Old Problems

By thinking logically, many times you will discover that the solution to a problem was right under your nose. For example, I recently went to the dentist. I was sure I would have to suffer when they took the X-rays. (I've always hated those little square X-ray plates they put in your mouth.) Lo and behold, they had a new kind! Someone had gotten the idea to take practicality, mix it with the old problem, and come up with a winner—sponge bite grips. Wonderful! No more pain. Too bad we've suffered all these years.

Giving It a New Twist

This type of invention involves taking an old product that may be past its market prime, and reinventing new uses for it. A perfect example is the bubble packs used to package breakable items being boxed for shipping. Some wise young father figured it made a perfect toy for his kid, since all kids (and we older kids, too) love to "pop" the bubbles. *Voila!* A new toy, and new inventor—and megabucks!

THE THREE BASIC TYPES OF INVENTION IDEAS

There are three basic ways to come up with ideas for inventions: (1) spontaneous ideas, (2) direct efforts to solve current problems, and (3) daily problem solving.

A *spontaneous* idea comes when you're sitting in the bathtub, and an idea for a wild new kid's game or toy pops into your mind. (Don't laugh; some of my better ideas for books take place in the bathtub.)

A *conscious, direct effort* to solve a problem occurs when you apply a detectivelike approach to the problems facing you.

Daily problem solving means working on a project or idea not directly related to another one, yet you discover "by accident" the solution to a problem that has been plaguing people for a long time. This was apparently the way rubber was invented.

At this point, you should find out where most of your creative ideas and inventions come from. Are they spontaneous, conscious efforts, or the results of daily problem solving? Take a minute to jot down a quick list. When you want to generate new ideas, concentrate on the method that works best for you.

Invention Idea Worksheet

Spontaneous	Conscious Effort	Problem Solving
Holster		
Aptih		
Headache		
Key sonic alarm		

WHY YOU SHOULD INVENT SOMETHING

This great country of ours would not be where it is today if not for inventors. Inventors like Benjamin Franklin (no relation to me, sorry to say) designed the Franklin stove, a printing press, the forerunner of the U.S. Postal Service, and many others. Closer to today, inventors like Steve Jobs and Steve Wozniak revolutionized the personal computer industry when they invented the prototype for the first Apple computer in their garage.

You should invent something because it does one or all of the following:

1. Fulfills a need in society.
2. Provides self-gratification.

3. Fulfills an economic need in society.
4. Makes you money.

At the risk of sounding mercenary, the last reason should probably be your first. I'm not saying you should be motivated by money above all else. But if money is not important enough, you won't do what's necessary to get a return on your investment—the marketing.

If you can fulfill a need, you will automatically make money. The golden rule of marketing is, "Find a need and fill it." To this rule I add, fill it correctly, with the proper invention.

WHERE DO INVENTIONS COME FROM?

Inventions are bred from ideas. Ideas come from ordinary people and extraordinary people like you. They come from people of all walks of life—professional workers, scientists, teachers, garbagemen, military personnel, housewives and househusbands—anyone and everyone.

The scenarios of where an idea can strike are numerous. In Chapter 4, we'll go through the thought processes of where you might look for invention ideas, and how to turn these ideas into a possible product.

First, however, we must focus on the basic setup of your invention business and the skills necessary to run it properly.

2

The Right Stuff: The Skills You Need

Every business requires certain skills essential to success. As an inventor, you, too, need skills in order to market your product properly.

Part V (Chapters 11–16) discusses the basic tools you will use to get your product to market. Your SUCCESS SKILLS, however, come first. Without these skills, the tools won't work.

There are three categories of skills you need: *creative, entrepreneurial, and marketing*. Let's examine each one.

CREATIVE SKILLS

If you can't create "a better mousetrap" or a new invention, you will have nothing to sell—and hence no money. Yet it would be presumptuous to assume you are creative all the time. Neither can we assume you know how to take your ideas and transform them into workable, useful products.

What are the creative skills necessary for success, and how can you attain them? In her book, *Creativity in Business,* Carol Kinsey Goman talks about a "creativity quotient," similar to an IQ. I suggest you take the short CQ quiz below, to help you determine the best way to create ideas.

1. When do I come up with the most ideas?
2. Where does this usually happen?
3. Do I create better by myself, or with others?
4. Does one idea trigger others, or are they separate?

For questions 1 and 2, list all the answers that come to mind. For example, you might list ideas you've gotten while showering, golfing, or driving. If your best ideas occur while showering at 8:00 A.M., list them together under "Showering." Do the same with other locations where you are creative, and note the time. List all ideas that were created by you alone, and with other people. You will begin to see a pattern develop showing where and when your better ideas come from.

If you don't already have an Idea Book (not the same as an Inventor's Journal or notebook), start one right away. A simple spiral notebook is best. List one idea per page. As you begin to flesh out the idea, your page will fill up. Name the product, idea, or concept at the top. When you add sketches, diagrams, and the like, transfer the information on the page to your Inventor's Journal. The notebook will take on intrinsic value for you, as it is the first step in protecting your invention (see Chapter 5).

ENTREPRENEURIAL SKILLS

While creative skills are the basis for developing products, being an inventor really means *running a business*. This involves inventing, protecting, packaging, distributing, promoting, and selling the product. You must develop these skills in order to learn how to run a small business each day. Hundreds of books have been written on the subject, some of which are published by the Small Business Administration (SBA), an agency of the federal government.

Entrepreneurial Traits

To start, let's test your "entrepreneurial quotient," as we did your creativity quotient. Answer the following questions as honestly as you can. Don't answer the way you *think* they should be answered; answer them with your feelings.

Are You a Self-Starter?

_____ I do things on my own. Nobody has to tell me to get going.

_____ If someone gets me started, I keep going all right.

_____ Easy does it. I don't put myself out until I have to.

How Do You Feel About Other People?

_____ I like people. I can get along with just about anybody.

_____ I have plenty of friends—I don't need anyone else.

_____ Most people irritate me.

Can You Lead Others?

_____ I can get most people to go along when I start something.

_____ I can give the orders if someone tells me what we should do.

_____ I let someone else get things moving. Then I go along if I feel like it.

Can You Take Responsibility?

_____ I like to take charge of things and see them through.

_____ I'll take over if I have to, but I'd rather let someone else be responsible.

_____ There's always some eager beaver around wanting to show how smart he [or she] is.

How Good an Organizer Are You?

_____ I like to have a plan before I start.

_____ I do all right unless things get confused. Then I quit.

_____ I take things as they come.

How Good a Worker Are You?

_____ I can keep going as long as I need to. I'll work hard for what I want.

_____ I work hard for a while, but when I've had it, I quit.

_____ Hard work doesn't get you anywhere.

Can You Make Decisions?
_____ I can make up my mind in a hurry if I have to.
_____ I can if I have plenty of time. I don't rush things.
_____ I don't like to be the one who decides things.

Can People Trust What You Say?
_____ Of course they can. I don't say what I don't mean.
_____ I try to be on the level most of the time.
_____ Why bother if no one knows the difference?

Can You Stick with It?
_____ I make up my mind to do something—nothing stops me.
_____ I usually finish what I start—if all goes right.
_____ If it doesn't go right, I quit. Why kill yourself?

How Good Is Your Health?
_____ I never run down!
_____ I have enough energy for most things I want to do.
_____ I run out of energy sooner than most.

Source: Checklist for Going into Business, SBA Management Aids, No. 2.016, U.S. Small Business Administration, 1985.

Count your check marks. If most of your checks are beside the first answers, you probably have what it takes to run a business. If not, you're likely to have trouble. Find a partner who's strong on your weak points. If there are too many checks next to the third answer, running a small business may not be right for you.

Bookkeeping Skills

While it is not necessary to be a bookkeeper yourself, you should know the basics about accounts receivable, accounts payable, inventory control, and cash flow. Payroll, general ledger, and credits and debits get too tough

Chapter 2 The Right Stuff: The Skills You Need   13

even for me, so I hire a part-time bookkeeper (an independent contractor) to keep me up to date. I have read, however, the basic SBA booklets on what goes into a company budget, and I strongly suggest you do also. (Check your local SBA office for any of their management aids booklets and business plans.)

MARKETING SKILLS

Marketing is everything you do to promote your business. It includes all forms of communication (writing, speaking, negotiating, selling), specific advertising and promotion techniques (mail-order advertising, direct-mail sales letters, cross-promotions, display and classified advertising), and public and community relations.

In the inventor's world, some marketing techniques work better than others. But you must be aware of all of them, and of what skills are needed to carry them out successfully.

Some of the basic skills you will need are:

1. Writing skills
2. Public speaking skills
3. Selling skills
4. Time-management skills
5. Networking skills
6. Negotiating skills
7. Research skills

Writing Skills

Basic writing skills are a must for an inventor. In addition to daily annotations in your notebook, you will be required to write letters to corporations, to the Patent and Trademark Office, to patent attorneys, to store buyers and sales reps, and to people in the various media.

Inventors are very technical and tend to write in a technical style. Most people cannot understand this kind of writing, however. It's best to learn to write businesslike letters in a simple easy-to-understand style. (See

Chapter 13 for some helpful hints.) There are many sources for learning basic writing skills.

Community College Classes. Every junior college and community college has basic writing classes that can help improve your prose. Check current class schedules, as well as credit and noncredit course offerings.

Books. There are many volumes of writers' books in libraries and bookstores. I recommend:

- *The Elements of Style,* by William Strunk, Jr., and E. B. White. New York.
- *The Associated Press Stylebook and Libel Manual.*
- *How to Write and Sell the 8 Easiest Article Types,* by Helene Schellenberg Barnhart.
- Any in the beginner's series from *Writer's Digest.*

Private Seminars. Private seminars on how to write are given every month all over the country. Check current issues of *Writer's Digest Magazine* and *The Writer Magazine* for listings of seminars near you.

Writers' Clubs. San Diego, California, has a good writers' club that meets once a month. Members critique each other's work and share techniques and skills. I'm sure there's a club in your area. Check *The Writer's Yellow Pages,* found in any library, for listings.

Speaking Skills

Good speaking skills are essential. Some of the marketing techniques for your invention may involve speeches, seminars, workshops, radio and television interviews, and others. If you lack speaking skills, then it's time to acquire them. There are many sources for learning speaking skills.

Speakers' Groups. Toastmasters International is one of the finest organizations around, especially for the beginner. Members guide you, step by step, through each aspect of speaking: appearance, presentation, diction, content, etc. To find the nearest chapter, contact the national organization and ask for the group nearest you: Toastmasters International, 23182 Arroyo Vista, Rancho Santa Margarita, CA 92688, (714) 858-8255.

Community Associations. Most community organizations, like Rotary, Kiwanis, and local Chambers of Commerce, have weekly breakfast or lunch meetings. Here's your chance to get up and brag about your invention for a few minutes. It's a great way to learn to speak in front of groups, and it's perfect for networking.

Community College Classes. Like writers' classes, most junior and community colleges offer speaking classes for credit or noncredit. Check your local junior or community college.

Books. While there are several good books on speaking skills and presentations, the one I like best is Leon Fletcher's, *How to Speak Like a Pro*.

Fletcher teaches how to structure and outline a speech. In writing down the structure, you automatically create an outline. This becomes the basis not only for a good speech, but for future articles, books, seminars, and press releases. (Since writing is just transferring spoken words to paper, you learn both skills together.)

Selling Skills

Selling is perhaps the most important—and most overlooked—skill. It doesn't do you any good to have the world's greatest invention if no one will buy it. A garage full of product does not pay the bills.

Most inventors don't know how to sell. If you do, congratulations! For you, this will be review. For those unfamiliar with sales techniques, the four-step approach to a sale is: Product Knowledge (about your invention), Finding the Buyer's Hot Button (probing), Answering Objections (trial closing), and Closing the Sale (asking for the order). (See Chapter 13 for a discussion of the AIDA selling technique.)

Books. It seems like there are more books on selling techniques than on any other subject. How do you find the good ones? Here are a few recommendations. These are the ones most salespeople consider outstanding.

Tom Hopkins' *How to Master the Art of Selling Anything* will give you an intensive course in salesmanship. Zig Ziglar's books and tapes are also good. A new book, Jim Cathcart's *Relationship Selling: How to Get and Keep Customers*, is particularly interesting. I like Jim's book, because I believe relationships are what selling is all about. This book gives you the basic skills, and translates them into human psychological methods.

Time Management Skills

Organizational skills are very important to the inventor. Because we wear many different hats, we must budget our time, and place priorities on those items giving us the biggest return. An excellent book is Stephanie Winston's *The Organized Executive*. Every inventor must have a Day Planner, in addition to an Inventor's Notebook and an Idea Book. A Day Planner is indispensable for helping you organize your time, and any of the many available kinds will do.

Networking Skills

Networking is the practice of meeting other people and using their connections to your benefit, while you give them connections to benefit them. It's the old "You scratch my back, I'll scratch yours." Networking is the current buzzword.

The ability to interact with people is very important to the inventor. The person who likes to stay in the shop and just create won't sell many products. If you are that kind of individual, it's best to get someone to do the marketing and selling for you. This book is for those who intend to do their own marketing, or at least a portion of it.

Here are several ways to get involved:

1. Form a mastermind group.
2. Join your industry association.
3. Form a mini leads club.
4. Join a community or civic group.

Mastermind Groups. A mastermind group is a small cadre of people within your industry or field who can share ideas and provide resources. While all members are inventors, each should be a specialist in a certain area of expertise, product material, or industry (such as toys, aerospace, etc.).

Get together informally to share ideas. While some members may be direct competitors, as long as you just talk about marketing and not actual product, it will help.

Industry Associations. If your project is electrical, join the Institute of Electrical and Electronics Engineers (IEEE). If you're working on hardware-related inventions, join the National Hardware Wholesalers' Association. To find industry associations, check Gale's *Encyclopedia of Associations* in your local library's reference section.

Leads Clubs. A leads club is an organization that meets once a week for the express purpose of exchanging potential prospects. While you don't want to sell your product just a few at a time, joining such a group might help with future volume leads.

Most formalized clubs charge a membership fee and have very strict rules, so I suggest you form your own. For example, you probably have a doctor and dentist, and maybe an accountant, banker, or lawyer. They should automatically be on your team—after all, they have businesses just like you. Why not meet once a week to exchange leads? Bring them a few names of people who may need their help, too. This is an example of referral at its best.

The rules of a mini leads club are:

1. Meet once a week for breakfast.
2. Bring two good leads to the meeting (prequalified, based on your knowledge).
3. Have only one member per industry in the group.
4. Limit the group to six.

Community and Civic Groups. Joining a community or civic group is a wonderful way to learn basic public speaking skills. However, the main benefit is networking with community leaders. Some of these people might have connections that can help you. You might end up speaking to them as a group, and find some buyers at the meeting.

Negotiating Skills

During your career, you will find it necessary to negotiate with salesmen, vendors, buyers, media people, corporations, and perhaps even the U.S. government. It is therefore essential to learn basic negotiation techniques. You don't want to give away the store. There are two main sources for learning negotiating skills.

Seminars. Take a course or seminar in negotiation tactics. One of the best is Karras Seminars in Los Angeles. It's well worth the money.

Books and Tapes. If your funds are limited, the Karras Institute offers its courses in book form and on audio cassettes. Roger Fisher's *Getting to Yes* is a classic in this field. Economics Press publishes *Negotiating Techniques,* and Nightingale-Conant has a tape series by Roger Dawson called *The Secrets of Power Negotiating.*

Role-play with some friends. You're the inventor, they're the corporation buyer. Have them be brutally critical of your invention. Then reverse the scenario. This will make it much easier when the real situation comes along.

Research Skills

You already know how to research for your invention. Now you need to learn what's necessary for market research.

In Chapter 11, we'll discuss basic market research, and how to pinpoint what we want to sell, who we want to sell to, and where we can find customers. In the meantime, get used to the idea of the library as your "fortress of solitude," and the reference librarian as your best friend.

If you think you'll need additional help, consider hiring a college student part-time who's majoring in marketing. Your project will give them a chance to learn from real-world experience, and it can be an inexpensive way for you to get valuable assistance.

LET'S GET STARTED RIGHT!

If, like most readers of this book, you intend to take the self-educated route and do your own marketing (the most practical, logical, and initially the safest way), I suggest the following approach:

1. Identify your strong skills.
2. Identify your weak skills.
3. Identify the skills you wish to learn.
4. Identify the skills you wish to delegate.
5. Start working on the skills in #3.

3

Getting Started Right: The Basics of Business

In any business, one must get started on the right track. Like a locomotive train bound from Los Angeles to New York, your business must follow a true path. There will be mountains and curves along the way that can slow you down. But the person who knows where he or she is going, and how to get there, will have a lot smoother ride.

The purpose of this book is to make the journey as smooth as possible. This chapter will help you get started.

There are many things you need to do to set up your business before you turn your ideas into marketable products. They are covered under the following headings and are shown as a flowchart in Figure 3.1.

1. Setting goals
2. Determining your type of business
3. Selecting a company name
4. Licensing and red tape
5. Checking up on trademarks
6. Providing stationery, letterhead, and envelopes
7. Getting business insurance
8. Learning the rules of the road

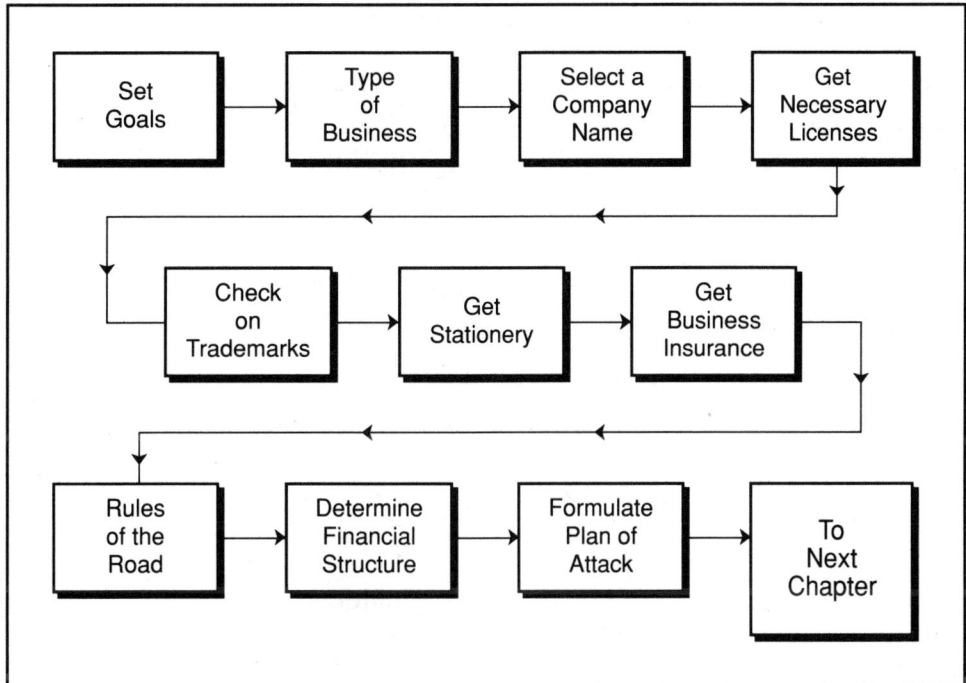

FIGURE 3.1 Getting Started Right: The Business Basics

9. Determining your financial structure
10. Formulating a plan of attack

SETTING GOALS

In addition to the traits mentioned in Chapter 1, you must set goals, both personal and business. These goals should be written. If you write them, they happen; if you just say them, they won't.

For goals to be realistic, you must be able to describe them with five essential qualifiers: short-term, measurable, exciting, big, and targeted.

Short-Term Goals

Goals that take too long to achieve are unreachable, and goals that are unreachable are useless. People don't pursue them for very long. They get bored and move on. To achieve success, make each of your goals short-term.

Measurable Goals

If you can't measure your goal, how do you know if you can achieve it? For example, if one of your goals is to sell to a corporation, set a time limit to see a specific number of companies during this period. If you reach this goal, you've measured success.

Exciting Goals

A goal must be exciting or there's no sense in setting it. You won't be happy, and you certainly won't be motivated to complete anything.

Big Goals

Goals must be big enough to provide a challenge. They must be within reach, but not too easy. If your goal is too far away, you won't even try. If it's too easy, why bother?

Targeted Goals

Your goals must be focused. This is the most important element. You cannot be all things to all people, nor can your invention. Design your product with specific target markets in mind. If additional markets are interested, great! That's gravy.

Now that you understand about goal setting, take a few minutes to answer the following questions. They should help you focus.

1. What are my personal financial goals for the next 6 and 12 months?
2. What are my business goals for the next 6 and 12 months?
3. Make a list of six potential inventions you want to work on in the next 6 months.

Now go back and rank them, according to the level of excitement you feel about each goal.

DETERMINING YOUR TYPE OF BUSINESS

In starting your business, you will have to decide the kind of organization you wish to have. There are four kinds to choose from:

1. Sole proprietorship
2. Partnership
3. Corporation
4. Subchapter S corporation

Usually, inventors are either sole proprietors or members of a partnership. These forms are easier and less expensive when just starting out. However, if you intend to form a corporation, get help from an attorney. (You can do it yourself, too. A good set of books to help you is the *Starting and Operating a Business* series by The Oasis Press.)

Sole Proprietorship

Definition: A business owned by only one person.

Advantages:

- Ease of formation.
- Sole ownership of profits.
- Control and decision making vested in one owner.
- Flexibility.
- Relative freedom from government control and special taxation.

Disadvantages:

- Unlimited liability.
- Unstable life of business (i.e., if owner should die).
- Less available capital, ordinarily, than in other types of business structures.
- Relatively difficult to obtain long-term financing.
- Relatively limited viewpoint and experience of owner.

Partnership

Definition: A business owned by two or more persons.
Advantages:

- Ease of formation.
- Direct rewards.
- Growth and performance facilitated.
- Flexibility.
- Relative freedom from government control and special taxes.

Disadvantages:

- Unlimited liability of at least one partner.
- Unstable life of business (if one owner should die).
- Relative difficulty in obtaining large sums of capital.
- Firm bound by acts of one partner as agent.
- Difficult to dispose of partnership interest.

Corporation

Definition: A legal entity distinct from those parties or individuals that own it.
Advantages:

- Stockholder liability limited to fixed investment amount.
- Ownership is readily transferable.
- Separate legal existence.
- Stable and relatively permanent existence.
- Ease of securing capital from many investor sources.
- Corporation can draw upon the expertise of many people.

Disadvantages:

- Activities limited by the charter and various laws.
- Possible manipulation of minority stockholders.
- Extensive government regulations.

- Extensive required local, state, and federal reports.
- Less incentive to employees if they don't share profits.
- Expense of forming the corporation.
- Double taxation.

Subchapter S Corporation

Definition: A business structure combining the advantages of a partnership with those of a corporation.

Advantages:

- No regular corporate tax.
- Long-term capital gains taxed directly to shareholders.
- Tax on unreasonable earnings generally will not apply.
- Can remove prior taxed earnings from a corporation in cash without current dividends tax.

Disadvantages:

- Cannot have more than fifteen shareholders.
- If it receives too much passive or foreign income, you can lose the subchapter S status.
- Foreign tax credit not allowed for foreign taxes paid by corporation.

Since most tax laws are very complex, talk with your attorney, tax consultant, or accountant before you proceed with either a corporation or subchapter S.

For further information on business structures, consult Antonio M. Olmi, *Selecting the Legal Structure of Your Firm*, SBA Management Aids, No. 6.004, U.S. Small Business Administration, 1985.

SELECTING A COMPANY NAME

Your company name is most important. In marketing, image is everything, and the name you give your company will either detract from or enhance its image.

Most entrepreneurs and inventors start with a company name composed of their own name with the addendum "and Associates." This is all right if you intend to stay small forever. If you plan on expanding and growing in the future, you should consider alternatives.

Take the time to consider a good name. Make a list of all the company names you admire. Then make a list of all the possible kinds of products you intend to work on (e.g., electrical, mechanical, toys, jewelry, etc.). Put word combinations together, and see how you like them. Test them on your friends and relatives.

For example, one of my former students manufactures jewelry. She intends to branch out into women's sportswear and beach clothing. The name she chose, Kiazi (pronounced Kee-ah-zee), is perfect. It gives the impression of sophistication, elegance, and high fashion.

Whatever you choose, it's OK to use your name with the addendum "Company," or "and Company." But avoid the word "Enterprises." It sounds amateurish.

You might consider using your product name for your company name—for example, Rite-Well Pen Company. The problem with this, however, is that it limits you to a specific product identification. If you intend to invent other products, you put yourself in a trap.

LICENSING AND RED TAPE

To be legal, you will probably need several different types of business licenses, all at various levels of government.

Fictitious Business Name Permit

If your company name is anything but your own, it would be a good idea to get a fictitious business name permit. This can be obtained from the local county clerk. (Even if you have "and Associates" with your name, play it safe. Some cities require a fictitious name permit, and some do not. Check with your city or county clerk.)

1. Check the register of names for any names that are similar to the one you are considering. This might confuse future customers.

2. Fill out the form, listing your fictitious name, actual name, address of business, phone, partners or owners, and other pertinent information.

3. Pay your fee. Where I live, the fee is $10 for the first name and $2 for each additional name. (If you fail to list all the names at the initial time of filing, you will have to refile a supplement later, and rerun another listing in the local paper.)

4. Run a small legal ad in the local paper in the space reserved for such notices. The requirement is usually four weeks, and the cost ranges between $25 and $55, depending on the publication. Check with your local city newspaper, and shop around for the best price.

Local Licenses

Most cities and or counties require you to get a business license. This is a simple process.

Go down to city hall, and get a business license permit form. Fill it out, pay the clerk, and take your receipt. The clerk's office will normally send your license by return mail. (You must renew the license every year.)

The fee is determined by the nature of your business and your estimated gross sales for the year. So be conservative. Don't say you plan to sell a million widgets. Bragging will cost you money.

State Licenses

Most state governments require you to have a special license for certain types of businesses. Check with your secretary of state to see if you need one.

If you intend to sell your invention to retail customers, you will need a resale number from your state. This permit allows you to act as the state's agent in collecting state sales taxes (if any).

When I went to get my resale number, I found out several things that should prove useful to you:

1. It takes 4 to 5 hours to get your license, so plan to spend all day. The number of businesses applying for permits far outweighs the examiners who interview you.

2. If you are more than 30 days away from selling your product, they will not issue you a resale number and will ask you to come back.

28   *Part I   In the Beginning*

3. You cannot get a resale number by mail. The licensing office requires an interview to establish the legitimacy of your business.

4. If the examiner feels you will gross a large sales volume each quarter, he or she may ask you to post a bond as a deposit against future sales taxes collected. They hold this money and don't pay you interest. So dress casually, and underestimate the amount of sales you intend to make.

5. You must pay sales taxes quarterly. Don't forget, this is *not* your money; you are only a collection agent for the state. If you miss your filing deadline, you can lose your permit and will have to start over again.

Federal Licenses

If your invention falls under federal guidelines for alcohol, firearms, tobacco, food, electronic appliances, and others, you may need special permits or licenses.

For example, electronic products emit certain frequencies. These products must be licensed by the Federal Communications Commission (FCC) and stamped "FCC Approved." Food and drug products must be manufactured and labeled according to standards set by the Food and Drug Administration (FDA).

Check with your industry association to determine whether you need a special permit. You can find a list of associations in Gale's *Encyclopedia of Associations,* located in the reference section of your library.

Special Licenses

You may need special licenses not covered above. For example, restaurants must have a city or county health permit to operate. Your invention may be similarly restricted. Check the local laws. You cannot afford to be surprised.

CHECKING UP ON TRADEMARKS

Many company names include a logotype or trademark as part of their identification. Before you file your business or product name, see if someone else already owns that trademark. Gale's publishes *Brands and Their Companies,* found in most county libraries.

In addition, consider trademark protection for your company name and products (see Chapter 5). Check on the scope of trademark protection for your name. A client recently found that her company name was only trademarked in California. (I knew something was wrong when she told me it only cost her $50 to file. Normally, trademark fees run $250 and above. Trademark searches cost extra.) It cost her an extra $175 to protect the name in the rest of the country, but it was well worth it.

PROVIDING STATIONERY, LETTERHEAD, AND ENVELOPES

Naturally, you realize the need for a professional image. So make your stationery, letterhead, and envelopes as professional looking as possible. You don't have to take time to research type styles, layouts, and design, but you should get the most for your money. Here's what I suggest:

1. Get six bids from shops that provide typesetting and printing for business products.

2. Tell each one the kind of company image you want, and the field of products you are involved in. (Don't spill the beans; just say, for instance, "I'm involved in the toy industry, what can you show me in design and type that says toys?")

3. It doesn't cost much more to print two colors than one. Two colors look more impressive. (To save money, print black ink on colored paper for a two-color look.)

4. Understand basic printing terms. Two-color means black plus one extra color. The primary printing colors are black, cyan (blue), yellow, and magenta (pink). (Red is a three-color process: black plus yellow plus magenta. A three-color printing job costs more than a two-color job.) PMS colors are premixed and you can pick them off a color chart. Prices vary. Ask questions, before you make a color decision.

5. Another way to save money is called a gang run. Your printing job is put at the end of someone else's that is printing the same colors as you are. The printer charges you for the print run and color, but not the preparation of the printing plates (called make-ready). If you tell your printer to gang run your job with a bigger job and you have the patience to wait, you'll save money.

6. There's one more way to save money on printing. Some printers have certain color days. On Monday, they print blue; Tuesday, red; and so on. By waiting until your color day, you save again on preparation costs.

GETTING BUSINESS INSURANCE

All businesses need insurance. You can be sued for almost anything these days—so protect yourself! These are the types you *must* have:

1. *Automobile insurance.* If you drive, you already have this type of insurance.
2. *General business insurance.* This type includes liability in your workshop, office, and home.
3. *Product liability insurance.* A client of mine found this one out the hard way. No major company would take her toy in their stores unless she had a $1-million liability insurance policy.
4. *Key man insurance.* This type of insurance keeps the company going while you recover from a serious illness. If you have a partnership, both partners must have it.

The SBA recommends the following insurance checklist. While you may not need all the types listed, they are good to know about. Check with your insurance company first.

1. *Fire insurance.* For inventory, buildings, office equipment, show equipment, vehicles, records, currency or securities, employee personal property, smoke and water damage.
2. *Floaters.* For theft, burglary, floods, windstorm, explosions, riots, vandalism, plate glass breakage, lightning, business interruption.
3. *Life and health insurance.* For key man, hospitalization and medical care, life insurance.
4. *Liability insurance.* For products, off-premises accidents of employees and nonemployees.

5. *Vehicle insurance.* For collision, bodily injury, theft, towing, property damage, and liability.

LEARNING THE RULES OF THE ROAD

Certain inventions may require more stringent regulations than those already mentioned. For example, if you invent toys, there are rigorous child safety regulations you must adhere to before your product reaches the market. Check with your industry association.

DETERMINING YOUR FINANCIAL STRUCTURE

Whether you are a sole proprietor, a member of a partnership, a corporation, or a subchapter S corporation, each has certain tax advantages. Do the following:

1. Determine when your fiscal year begins and ends.
2. Decide whether your records and books will be set up on an accrual or cash method of accounting.
3. Outline your personal and business financial statements.

Personal Financial Statement

Because you, like most inventors, are likely to be a one-person operation, your initial personal financial statement may be the same as your business statement. (Keep them separate.)

You will need some type of statement to show bankers, venture capitalists, landlords, and others. Let's start with a personal financial statement.

This is nothing more than a simple listing of your personal assets and liabilities (what you own and what you owe). The difference is called your net worth. A sample is shown in Figure 3.2.

Personal Financial Statement

I Own		I Owe	
Cash		Current household bills	$ _____
Bank accounts	$ _____	Installment contracts	_____
Other	_____	Car	$ _____
Securities—quick-sale value	_____	Appliances	_____
Real estate—quick-sale value	_____	Personal loans	_____
Furniture—quick-sale	_____	Real estate mortgage	_____
Car	_____	Other	_____
Cash-value life insurance	_____	(describe _____	
Savings bonds	_____	_____	
Other assets	_____	_____)	
Receivables	_____	Other loans	
Total	$ _____	(describe _____	
I own	$ _____	_____	
I owe	$ _____	_____)	
My net worth is	$ _____	**Total**	$ _____

FIGURE 3.2 Personal Financial Statement

Business Financial Statement

In order for your business to be healthy and thrive, it will need a good positive cash flow. Like most start-ups (or expansions), you might need a business loan. Chapter 6 discusses the potential sources you can possibly tap for capital.

How much money you need, what goes into a budget, and other financial matters are discussed in Chapter 7. For now, you'll need the following tools: a Projected Statement of Sales and Expenses for One Year (Figure 3.3); an Estimated Cash Forecast (Figure 3.4); and a Current Balance Sheet (Figure 3.5).

Projected Statement of Sales and Expenses for One Year

Total Jan Feb Mar Apr May Jun Jul Aug Sep Oct Nov Dec

A. Net Sales

B. Cost of Goods Sold

  1. Raw Materials

  2. Direct Labor

  3. Manufacturing Overhead

      Indirect Labor

      Factory Heat, Light and Power

      Insurance and Taxes

      Depreciation

C. Gross Margin (Subtract B from A)

D. Selling and Administrative Expenses

  4. Salaries and Commissions

  5. Advertising Expenses

  6. Miscellaneous Expenses

E. Net Operating Profit (Subtract D from C)

F. Interest Expenses

G. Net Profit before Taxes (Subtract F from E)

H. Estimated Income Tax

I. Net Profit after Income Tax (Subtract H from G)

FIGURE 3.3 Projected Statement of Sales and Expenses for One Year

34   Part I In the Beginning

Estimated Cash Forecast

	Jan	Feb	Mar	Apr	May	Jun	Jul	Aug	Sep	Oct	Nov	Dec
(1) Cash in Bank (Start of Month)												
(2) Petty Cash (Start of Month)												
(3) Total Cash (add (1) and (2))												
(4) Expected Accounts Receivable												
(5) Other Money Expected												
(6) Total Receipts (add (4) and (5))												
(7) Total Cash and Receipts (add (3) and (6))												
(8) All Disbursements (for month)												
(9) Cash Balance at End of Month in Bank Account and Petty Cash (subtract (8) from (7))*												

*This balance is your starting cash balance for the next month.

Source: Business Plan for Small Manufacturers, SBA Management Aids, No. 2.007, U.S. Small Business Administration, 1987.

FIGURE 3.4   Cash Flow Forecast

Expected Sales and Expense Figures.   To determine whether your business can survive initially in the marketplace, estimate your sales and expenses for 12 months (Figure 3.3).

Cash Flow Figures.   Estimates of future sales will not pay an inventor's bills. Cash must flow at the proper times if bills are to be paid and a profit is to be realized at year's end. Prepare an estimated cash flow statement like the one in Figure 3.4.

Current Balance Sheet.   A balance sheet shows the financial conditions of a business as of a certain date. It lists what the business has, what it

Chapter 3 Getting Started Right: The Basics of Business 35

```
                    Current Balance Sheet
                             For
                    _____
                      (name of your company)
                            as of
                    _____
                             (date)

Assets                              Liabilities

Current Assets                      Current Liabilities

Cash                    $ _____   Accounts Payable    $ _____

Accounts Receivable     $ _____   Accrued Expenses      _____

Inventory                 _____   Short Term Loans      _____

Fixed Assets                        Fixed Liabilities

Land                    $ _____   Long Term Loan      $ _____

Building      $ _____             Mortgage              _____

Equipment       _____

  Total         _____             Net Worth           $ _____

Less
Depreciation            $ _____

Total                     _____                       $ _____
```

Source: Business Plan for Small Manufacturers, SBA Management Aids, No. 2.007, U.S. Small Business Administration, 1987.

FIGURE 3.5 Current Balance Sheet

owes, and the investment of the owner. It lets you see at a glance your assets and liabilities. Use the blanks in Figure 3.5 to draw up a balance sheet for your company.

FORMULATING A PLAN OF ATTACK

What I have been helping you set up is the starting point for a business plan. You cannot achieve success without one. You need the plan for obtaining capital, outlining your distribution and marketing strategies, and a whole lot more.

The basic points to cover in a business plan for inventors are:

1. A description of your business, including:
    Definition of business field you are in
    Company history
    Personnel
2. A marketing plan, including the following:
    Market area
    Competitors
    Distribution
    Market trends
    Market share
    Sales volume
3. A production plan, including:
    Manufacturing operations
    Raw materials
    Equipment
    Labor skills
    Space
    Overhead
4. A budget plan, including:
    Expected sales and expense figures
    Cash flow figures
    Current balance sheet figures
5. An organizational outline (company structure)
6. The product or products you intend to invent
7. Any supporting data

One of the most outstanding books on business plans is Joseph R. Mancuso's *How to Write a Winning Business Plan.* Joe is the founder of The Center for Entrepreneurial Management (CEM), a nonprofit membership association for entrepreneurs.

PART II

I'VE GOT AN IDEA!

4

From Idea to Product

In 1899, the Director of the U.S. Patent Office told President McKinley to get rid of the office, because "everything that can be invented, has been invented."

THE DEFINITION OF AN IDEA

Without an idea, there's no product. Without a product, there's obviously nothing to sell. Let's use the word *idea* as a benchmark to measure whether we should turn one into an invention.

IDEA = Innovation, Desire, Enthusiasm, Action

1. *Innovation.* To be marketable, your idea must be new and innovative. It can't be just a variation on the tried and true, because people will simply not buy enough to make it worth your while.
2. *Desire.* You must create within your target audience a burning desire to have your product. If they feel emotionally attached to it and can't live without it, they'll buy it. This emotion can stem from a real need, or you can convince your potential customers that there is a need.

The difference between "want" and "need" is very important here. To want is to be without that which gives comfort or is desired; to need is to be without that which is *essential* to existence or our purposes. What we *want* meets artificial desires; what we *need* satisfies real requirements. You should strive to find a product that satisfies a need.

3. *Enthusiasm.* You can do the best job selling your invention if you are enthusiastic about it. The saying "Enthusiasm breeds enthusiasm" is true. Make your target market *so enthusiastic* about your invention that they scream, "I've just got to have it!"

4. *Action.* Your idea is no good if it stays in the idea stage forever. Many an invention that could help humankind has lain dormant for lack of an inventor's sense of direction and purpose. You need an action plan that includes all three of the above criteria.

WHAT'S THE IDEAL PRODUCT?

There are four different types of products. Some are difficult to market; others are easier. (There is really no such thing as an easy sell. Even fads have to have marketing.)

1. Long life + limited customer potential = too costly, no resale value (e.g., an infallible jet airplane).
2. Short life + unlimited customer potential = good resale value (e.g., toothpaste).
3. Short life + limited customer potential = fad (e.g., Wacky Wallwalker).
4. Long life + unlimited customer potential = good product group (e.g., a dishwasher and other appliances).

As you can see, if your invention had repeat sales potential and everyone constantly needed it (customer potential), then you'd have the ideal product. That's why Proctor and Gamble and others keep "reinventing" laundry detergents and toothpastes. These staples are with us forever. In order to grab better market share, the top two or three companies keep extending their product line, or developing so-called new brands. But the products are really still basically the same.

WHERE DO YOU FIND IDEAS?

The first consideration is you. What do you like to do most? Make a list of what tasks, jobs, and daily activities you do that make you happy. If you can develop a creative idea from this list and turn it into an invention, your natural enthusiasm will project in your marketing. (See the above criteria again.)

Here's a partial list of places to look and ways to develop invention and product ideas:

Newspapers
Trade magazines
Television
Radio
Trade shows
Industry associations
Visit stores
Talk to people
Libraries
Take an old product, give it a new twist

In addition, the SBA's *Finding a New Product for Your Company*, Management Aid No. 2.006, suggests several sources for new products. Any one of these might be an excellent starting point for developing your own ideas:

Government-owned patents
Private patents
Large corporations
Inventors' shows
Commercial banks
Small business investment companies
Licensing brokers
Foreign licenses
New product advertising

Keep track of your ideas in your Idea Book, as described in Chapter 2.

HOW DO YOU CREATE IDEAS FROM THESE SOURCES?

Ask yourself working questions. Looking in the newspaper, for example, you see a story on child-care centers that are understaffed. Your working question might be: What could I invent that would make their jobs easier?

Another way is to take a current product on the market and change it. By modifying the shape, size, or structure, you develop a whole new product that might be better for the marketplace. Some working questions to ask are: Can I make it fatter or thinner? Larger or smaller? Heavier or lighter? Should I reverse it? Rearrange it? Adjust it? The mind map in Figure 4.1 gives you a starting point for idea generation. Draw your own mind map, and see how well it works.

ASK YOURSELF TOUGH QUESTIONS ABOUT EACH IDEA

Tough questions will help you weed out losing ideas before you commit additional time and resources. Listed below are questions you can use to screen most new product or service ideas. If you can't answer these questions, you probably need to investigate further:

1. Is there a genuine need for the product?
2. Is the need substantial enough to support a profitable business?
3. Do competitors currently offer similar products?
4. If "yes," does your idea offer distinctive advantages and customer benefits the competition's don't?
5. Is the product feasible to produce?
6. Is the product legal?
7. Is it safe?
8. Is the product a durable good, and can it be easily serviced? Who will service it?
9. Are the investment costs required to develop, produce, and market the product reasonable and within your financial realities?

Chapter 4 From Idea to Product   43

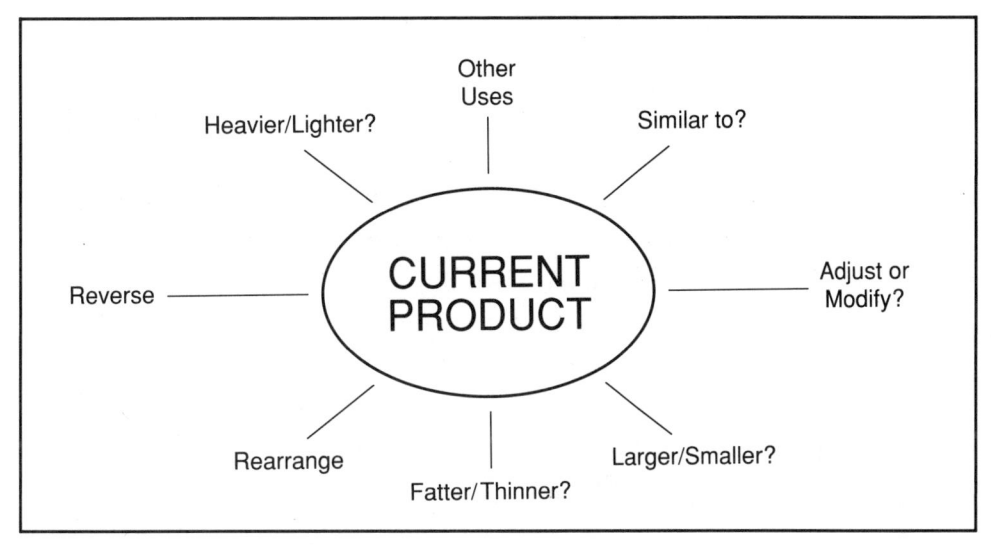

Mind map diagram: CURRENT PRODUCT in center, with branches to: Other Uses, Similar to?, Adjust or Modify?, Larger/Smaller?, Fatter/Thinner?, Rearrange, Reverse, Heavier/Lighter?

Source: Charles L. Martin, *Starting Your New Business,* Los Altos, CA: Crisp Publications, Inc., 1988. Used with permission.

FIGURE 4.1 Mind Map for Idea Generation

10. Is the "pay-back period" fast enough to allow you to stay in business?
11. Can the product be expanded into a line of similar or compatible items later, if the original was successful?
12. Are the needed raw materials and supplies readily available?

PRODUCT EVALUATION

Here are some additional questions to help rate your idea or product:

1. Does the product have any environmental impact?
2. What is the broad potential of the market?
3. How long is the life cycle of the product?
4. Is it functionally possible?
5. Will users of the product need training?
6. Are the start-up and tooling costs realistic?
7. Is the retail cost high enough to make a profit?
8. How will the distribution be handled?
9. Do you have proprietary protection? (See Chapter 5.)

CHECKING IDEAS FOR NEWNESS

1. *Check out the stores.* See if there are similar products already being marketed.
2. *Check with sales people at stores.* Don't give away your idea. Just ask, "Have you got X?" (X is a generic description of your product.)
3. *Check manufacturers' catalogs.* All manufacturers have some kind of catalog, which may include product descriptions and photographs. Compare these products with your ideas.
4. *Look through the Thomas Register of Manufacturers.* This can be found in the reference section of your local library. Find the names of manufacturers with similar products. Write to them for their catalogs and product descriptions.
5. *Check trade magazines. Standard Rate and Data Service* publishes a 13-volume set of directories, one of which is for trade magazines. Look up the SIC (Standard Industrial Classification) group section for your product industry. Then write to the top circulating trade magazines and ask for the latest issues, along with the year-end review issue. These will give you an idea of what's out there, and what the market (i.e., the editors) is talking about.

HOW WILL I KNOW IF THE IDEA WILL WORK?

This is the $60-million question, and one that nobody can accurately predict. However, with basic market research and planning, you can leverage the chances in your favor.

You need to do a *feasibility study* on your idea. Ask yourself these questions:

1. Is it feasible to market?
2. Is it feasible to make?
3. Are there enough customers?
4. Will it be profitable?
5. Is it a repeat item (fad or commodity)?

Is It Feasible to Market?

Most people would ask first if it's feasible to make, then market. I do it the opposite way. If I can't sell it, why bother trying to make it? All your energy must be marketing-driven in order to succeed (see Chapter 11).

Is It Feasible to Make?

Once you have determined that a potential market exists and that you have selling skills, you need to analyze whether you can make the product.

Are There Enough Customers?

If your market is too narrow, you will have to price your product high enough to make it worthwhile. This may be too high.

Will It Be Profitable?

You must figure out cost ratios and profit margins *before* you invest any real money. To do otherwise guarantees failure.

Is It a Repeat Item?

If your invention is considered a fad, you may expend energy and money on a quick thrill. Better to come up with an item that has repeat buyers, or at least a large enough universe so that everyone will need one, if only just once. Wacky Wallwalkers (those sticky spiders you throw against the wall), don't come along every day.

DEVELOPING YOUR PROTOTYPE

Once you've finished your idea analysis, it's time to develop a prototype. Notice, I did not say get tooling, make the product, get inventory, or anything else that people normally do. Don't jump the gun. We haven't even done a complete market or customer survey yet. (We will, in Chapter 11.)

A prototype has three purposes:

1. To demonstrate function.
2. To determine manufacturing requirements.
3. To be used in sales, marketing, and for investment purposes.

There are several ways to make a prototype. The easiest way is an artist's rendition. This is inexpensive and can be done rather quickly. Another way is to go to a plastics factory and have them develop (after signing a nondisclosure statement) a simple scale model. Also try a sheetmetal specialist, or a cabinetmaker. The point is, don't spend a lot of money. You want something that gives a rough idea of what the product will look like.

TEST-MARKETING

After you have the prototype finished, it's time to test-market your product in the real world. Ask the analysis questions about your customers, product, and competition to help you decide on product specifications (see Chapter 11). Now comes the real test. We put theory and research into actual testing—will someone give us an order?

Here are several ways for you to test your prototype:

1. *Ask other people's opinions.* Conduct an opinion survey by making a questionnaire. Show people the prototype and say, "We're thinking of putting this on the market. What do you think?" You will hear some very interesting comments.

I did this for a potential client once. He had a ski-locking device that he thought would sell like hotcakes. He had purchased the patent from another inventor. My kids lived in Mammoth Lakes ski area at the time. With survey and clipboard in hand, I proceeded to question skiers just down from the mountain. My 4 hours in the cold, with comments like, "Nah, I have a bike chain and lock that's good enough" convinced me I didn't have a very sellable product.

You can also try using a focus group. A focus group is 6–12 people in a room who evaluate your product or idea. You show them your prototype

Chapter 4   From Idea to Product   47

and let them "pick it apart." This is as close to real buyers as you're likely to get. One note of caution: Strangers work best. Your friends are prejudiced and won't give you an honest opinion.

   2. *Display the prototype at trade shows.* Trade shows get very sophisticated buyers; they know what will sell and what won't. You'll know quickly if you've got a winner or a loser (see Chapter 14).

   3. *Ask a store buyer.* Department stores have regular buying hours. Make an appointment to see a section buyer who deals with your particular product type. Ask her or him to critique the product. If it's good, you may get lucky and walk away with an order. At worst, you'll get valuable criticism.

   4. *Write to companies and ask for their opinions.* Be careful with this one—you could get burned. Make sure your product is protected with a nondisclosure agreement. Do this only if you intend to sell out to a corporation (see Chapter 8).

   5. *Do a small sample run.* You could do a small sample run, and place your product in a one-store test on consignment. If it sells, great! However, make sure the storeowner does not give you any more leverage than other products. If he is pushing your product too much, it's not a fair test.

ON TO THE NEXT STEP

Chapter 5 describes the ins and outs of protecting your invention. While this is very important information, I don't want you to get paranoid. Just protect yourself and you'll win the game.

5

Protecting Your Invention

There are four stages of invention protection: the untried and unsearched stage, the searched and reported patentable stage, the patent applied-for stage, and the patented product stage. Whatever stage you are in now, your ultimate goal is a full patent.

WHY YOU NEED PROTECTION

Your invention is important—it's part of you. You've put your heart and guts into creating it, designing it, making a prototype, and will want to test-market it. Taking this product into the marketplace without protection is like a parent taking a child out into the real world without any immunization. The child would probably get sick and might even die.

Treat your invention as you would your own child. Nurture it, care for it, and protect it from the germs that would do it harm.

In saying this, my purpose is to make you extremely cautious, not paranoid. Don't spend every moment looking over your shoulder for thieves in the night. Just remember: To be forewarned is to be forearmed. You'll sleep better, knowing that you've used every legal means to keep your invention yours.

But also be aware of this caveat: No protection in the world will stop someone from trying to rip you off. It just gives you the legal means to fight them in court. This is important enough to repeat. *Protection does not stop someone from making a knock-off of your invention* and trying to sell it in the same marketplace. It simply helps you fight legally to force them to "cease and desist." Sometimes even the mere presence of protection in the form of phrases like "Patent pending" or "Copyright, my invention, 1986" is enough to force a potential thief to think twice.

WHO CAN GET PROTECTION

Any inventor can get some type of protection for his or her invention, with certain restrictions. These restrictions have to do with timing, usefulness of the invention, and a host of other criteria. (See the section on patents later in the chapter.)

The more sophisticated the protection you want, the more you'd better be sure you have your act together. If you intend to patent your invention, for example, you'll need to have all your information put in the correct form—all your i's dotted and your t's crossed. This is why you may want some help.

HOW TO GET PROTECTION

There are two ways to get protection: by doing it yourself, or by using an attorney or small business consultant. In doing it yourself, you put the entire burden—filing applications correctly and on time, knowing exactly who to send forms to and how, and the myriad other instructions—on you.

You may not have the time, energy, or knowledge to do it all yourself. I pay other people to help me out. But I do have a general working knowledge of the different types of protection. How each one applies to an invention I may be working on is vital to know.

If you intend to use a small business consultant or an attorney, make sure you get one well versed in all the aspects of legal protection. Get three or four good consultants or attorneys, and discuss with each one your needs, your budget, and what you want them to do for you. Then pick the best one, not necessarily the cheapest, and go with that person. (For a list of registered patent attorneys, write to the Patent and Trademark Office, U.S. Department of Commerce, Washington, DC, 20231. For a list of business consultants, call your local SBA office for recommendations.)

THE THREE KEYS TO PROTECTING YOUR INVENTION

There are basically three key points you should keep in mind when looking for ways to protect your invention: keep quiet, maintain good records, and look for ways to improve your invention.

Keep Quiet

An old World War II expression is: "Loose lips sink ships." Tell no one about your invention until you have some form of protection. Have all people who see your invention, or who even discuss it with you, sign a simple nondisclosure agreement saying they will not divulge the idea to anyone else. See Figure 5.1 for an example. (Note that this sample nondisclosure agreement is not a legal document and should not be taken for one. It is a composite of many such agreements, used solely for the purpose of illustration here. To be on the safe side, check with an attorney.)

Maintain Good Records

Start with your Idea Book and continue with your Inventor's Journal on your project (one journal per project). The journal is a notebook for recording what you do and when you do it. Should you have to defend yourself or your invention in court, you will have written dated records.

Look for Ways to Improve Your Invention

Continue to work on your invention on a regular basis. Always try to update and improve it. Check competition and other patent searches to see if anyone else is trying to improve on your idea. You get there first, you win!

> Enclosed is a description, and other materials, of my idea for a [insert name of product]. After your review, I understand you will send me a written evaluation of the potential for my idea. You have 30 days in which to evaluate my idea. If, at the end of that time period, you find no value in the idea, you will return all materials to me by registered mail. You will not disclose the idea to anyone for at least 1 year from date of this agreement. If during the 1 year period you wish to discuss the idea further, or with other interested parties, you will contact me for permission in writing *first*.
>
> If you wish to participate in the marketing of the idea, I understand you will contact me in the future to arrange a mutually satisfactory royalty payment plan.
>
> (signature) _____ Date:
>
> Witness: _____ Date:
>
> Inventor: _____ Date:
>
> Witness: _____ Date:

FIGURE 5.1 Sample Nondisclosure Agreement

TYPES OF PROTECTION

There are two main types of protection: informal (nonlegal) and formal (legal). The formal or legal types are the most commonly used by inventors. Of the legal types, there are three kinds: trademarks, patents, and copyrights. We will explore each kind in detail.

Informal Protection

Many entrepreneurs and inventors have heard that they can protect themselves several ways without spending any money or using formal legal means. These informal protections are just that: nonlegal means by which inventors try to establish precedence and, hopefully, stop others from encroaching on their territory.

The sad fact is, unless these devices are legal entities, they will not stand up in court. I'm not saying not to use them; it's good to have them in your arsenal *along with* actual legal devices. Additional documentation to show when you invented something *may* sway a judge in your favor, but this is highly speculative.

52   Part II   I've Got an Idea!

One of the most common nonlegal types of protection is called self-registration of copyright. You send yourself a document by registered mail, and don't open it. The postmark shows date of inception of an idea or invention. This is *not* a legal document and will not stand up in court. Don't fall for this.

Another misconception is that you can use copyright law to protect an invention at a lower cost than if you were to get a patent. *Not true!* Copyright law pertains to certain aspects of inventions and patents to others. (See the sections that follow.)

The only informal device that will probably help you if you have to go to court is your Inventor's Journal. This should be a permanently bound, page-by-page description of your invention through its various stages. Each page must be dated and witnessed, with the phrase, "This device was explained to me and I understand it" signed by the witness. Each page should also have a sketch of the invention at each stage, a description of the idea, the function of the device, and a description of any experiments performed on the invention. Each time you make a change, make a new page.

Formal Protection

Formal protection refers to the trademark, patent, and copyright laws. Each kind is different and is used for specific purposes. Trademarks and patents are granted through the Patent and Trademark Office in Washington, DC. Copyrights are registered through the Register of Copyrights, Copyright Office, Library of Congress, Washington, DC.

TRADEMARKS

To make this section and subsequent sections easier for you, I have designed them in question-and-answer format. If your question is not answered, contact the appropriate agency at the number listed at the end of each section.

What Is a Trademark?

A trademark may be a word, symbol, design, combination of word and design, slogan, or a distinctive sound that identifies and distinguishes the goods or services of one party from those of another.

How Does It Differ from a Service Mark?

Some companies use the trademark, others use what's called a service mark. A trademark is used on a product or its package, and a service mark is used in advertising to identify an owner's services.

How Long Does a Trademark Last?

The length of time you can have a federal trademark registration is 10 years, with 10-year renewal terms. The rights to the trademark last forever, if it continues to be used in the manner in which it was registered (You must file an affidavit stating you are using the trademark between the fifth and sixth years.)

Do I Need to File with the U.S. Government?

It's not necessary to file with the U.S. government for a federal registration of your trademark, but it's a good idea. You still get the same protection without registration. But you get added benefits, like the right to sue in federal court. In other words, cover your assets!

When Can I File?

If you do decide to file for registration, you must first show you used the mark on your packages, or in your service, during the normal course of business.

How Do I Register My Trademark?

To file for registration, you must file a written application with the PTO in Washington, DC. There are four required items you must submit:

1. A completed application form (see Figure 5.2).
2. A drawing of the mark.
3. Five specimens showing actual use of the mark in connection with goods or services.
4. A filing fee.

54 Part II I've Got an Idea!

| TRADEMARK APPLICATION, PRINCIPAL REGISTER, WITH DECLARATION (Individual) | MARK *(identify the mark)* |
| | CLASS NO. *(if known)* |

TO THE COMMISSIONER OF PATENTS AND TRADEMARKS:

NAME OF APPLICANT, AND BUSINESS TRADE NAME, IF ANY

BUSINESS ADDRESS

RESIDENCE ADDRESS

CITIZENSHIP OF APPLICANT

The above identified applicant has adopted and is using the trademark shown in the accompanying drawing for the following goods: _____

and requests that said mark be registered in the United States Patent and Trademark Office on the Principal Register established by the Act of July 5, 1946.

The trademark was first used on the goods on _____ ; was first used on the goods in
                                                (date)
_____ commerce on _____ ; and is now in use in such
  (type of commerce)   *(date)*
commerce.

The mark is used by applying it to _____

and five specimens showing the mark as actually used are presented herewith.

_____
(name of applicant)
being hereby warned that willful false statements and the like so made are punishable by fine or imprisonment, or both, under Section 1001 of Title 18 of the United States Code and that such willful false statements may jeopardize the validity of the application or any registration resulting therefrom, declares that he/she believes himself/herself to be the owner of the trademark sought to be registered; to the best of his/her knowledge and belief no other person, firm, corporation, or association has the right to use said mark in commerce, either in the identical form or in such near resemblance thereto as may be likely, when applied to the goods of such other person, to cause confusion, or to cause mistake, or to deceive; the facts set forth in this application are true; and all statements made of his/her own knowledge are true and all statements made on information and belief are believed to be true.

_____
(signature of applicant)

_____
(date)

FORM PTO-1476FB (REV. 4-87)   U.S. DEPARTMENT OF COMMERCE/Patent and Trademark Office

Source: U.S. Government Purchasing and Sales Directory, Small Business Administration, 1984.

FIGURE 5.2 Trademark Application Form

Chapter 5 Protecting Your Invention   55

If you meet these requirements, your product or service is assigned a serial number, and you get a receipt.

Your application is then sent to an examining attorney to check for suitability. The examiner will issue any objection within 3 months. You then have 6 months to respond to the objection. If your application is refused again after you respond, you can appeal to the Trademark Trial and Appeal Board. If they accept, it is published in the *Trademark Official Gazette*. Anyone can then oppose it within 30 days. If someone objects, it goes to the Appeal Board. If no one objects, it is issued within 12 weeks of the publication. This process is shown in a flowchart in Figure 5.3.

What Symbols Do I Use, and When?

Once you get a federal registration, you may use the ® symbol, or the phrase "Registered in U.S. Patent and Trademark Office" or "Reg. U.S. Pat. & Tm. Off." Even if you don't intend to register, or before your application is granted, you can use the ™ or ℠ symbols.

Should I Do a Trademark Search, and Where?

It would be a very good idea to do one. One of the grounds for dismissal is that your trademark may look too much like someone else's. A trademark search can possibly prevent this. An attorney can do it for you, or you can use the Trademark Search Library in Washington, DC. You can also hire a trademark search firm, but be careful. There are sharks everywhere. The cost should not be too excessive, $90–$300 for a preliminary search.

There is also an excellent *Brands and Their Companies* directory published by Gale Research Corp. (Detroit, 1993) which lists previously trademarked names. Check your county library's reference section.

How Can I File?

Do it yourself, or hire an attorney to help you.

How Much Will a Trademark Cost?

The fee to the PTO is $210 for each class of goods or services for which an application is made. You send a check or money order in advance. For

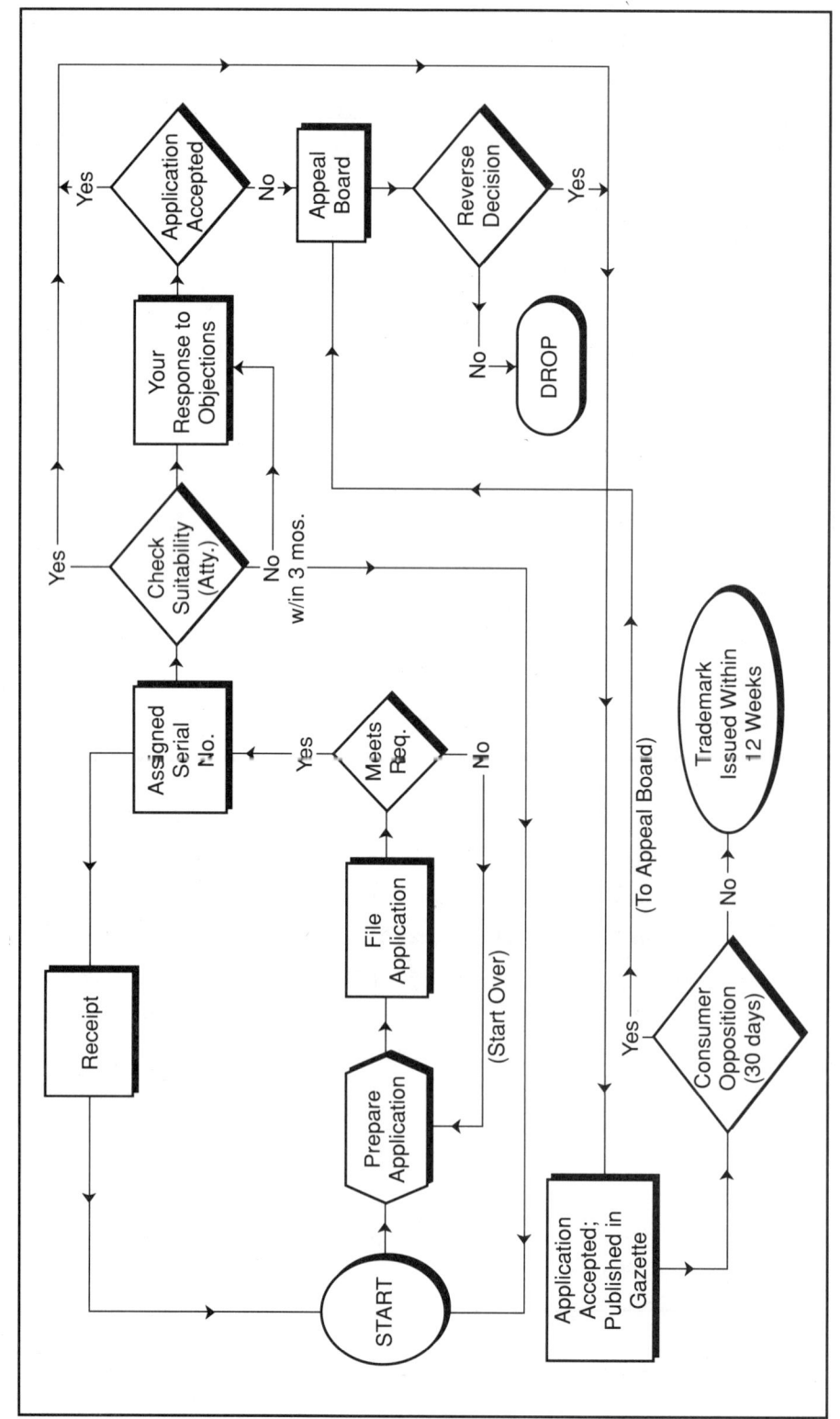

Source: Basic Facts About Trademarks, U.S. Department of Commerce, Patent and Trademark Office, Washington, DC.
FIGURE 5.3 The Trademark Process

further information, write to the Commissioner of Patents and Trademarks, Patent and Trademark Office, U.S. Department of Commerce, Washington, DC 20231. For general information, call (703) 557-INFO.

PATENTS

What Is a Patent?

A patent is the right granted to an inventor by the government to make or use a particular invention for a term of years without competition by other users of the same thing.

The actual article of patenting is called a Letter of Patent or Letters Patent. It is a document issued by the U.S. government's Patent and Trademark Office in Washington, DC. It is a very complicated process and should only be done by a registered patent attorney. The process is shown in a flowchart in Figure 5.4.

Patents can be granted for five different entities: designs, structures, processes, materials or combinations of materials, and living cells or combinations of cells.

There are also five parts to the patent application:

1. Petition: a formal request for a patent.
2. Specifications: written description of your invention.
3. Claims: description of the protection you seek.
4. Drawings: graphics and technical details.
5. Oaths: proof that you are the first and true inventor.

Of the above, #3, claims, is the most important and the most complicated. This part must be written in a formalized style. I advise getting the help of an attorney well versed in patent law. Interview four or five before making a decision. Fees are all over the place, from $500 to $5000!

How Long Does a Patent Last?

A patent is granted for 17 years from the date of issuance, subject to payment of maintenance fees due at 3.5, 7.5, and 11.5 years after issuance, and

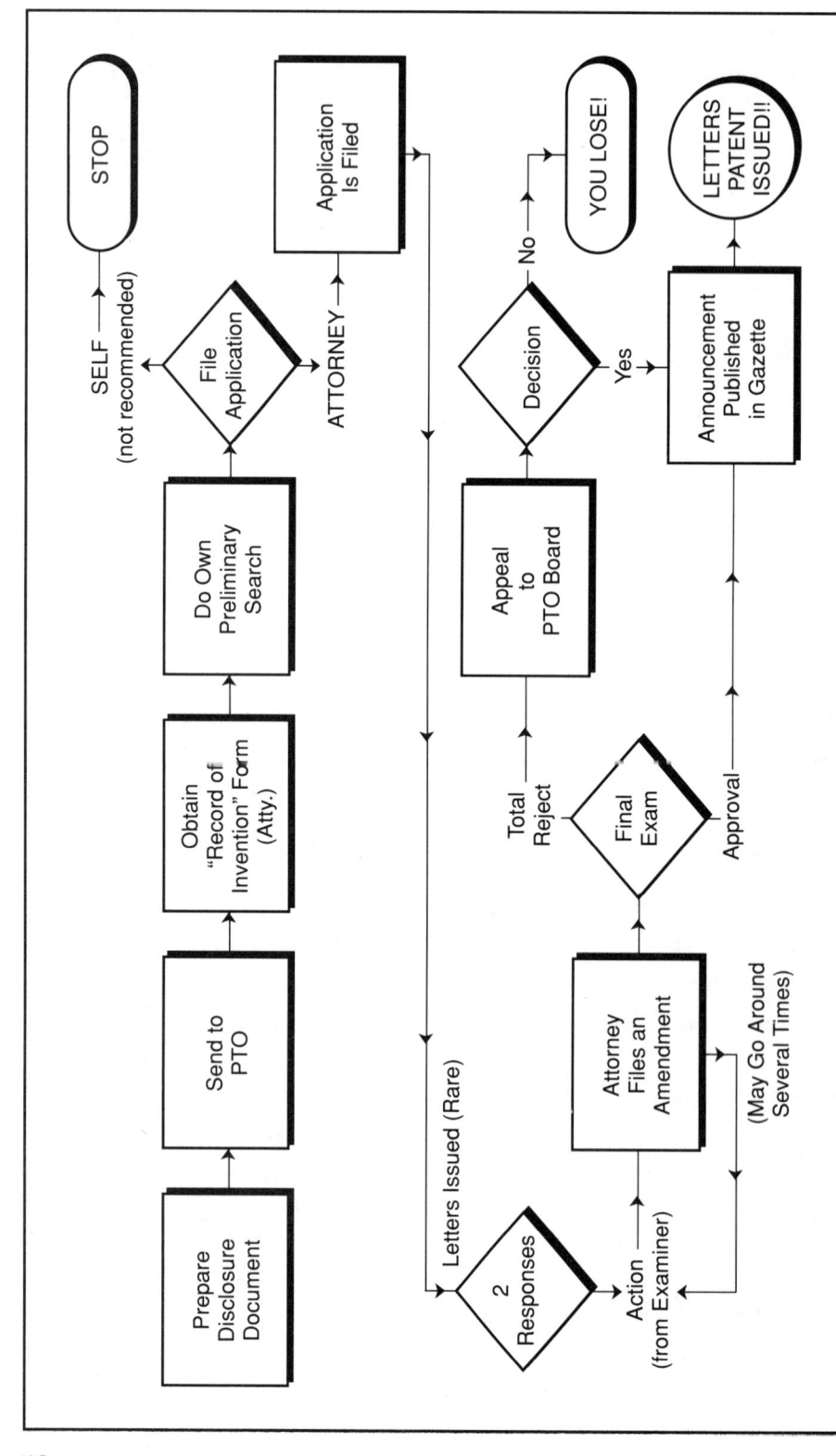

FIGURE 5.4 The Patent Process

Chapter 5 Protecting Your Invention   59

it cannot be renewed. After 17 years, your invention becomes part of the public domain, and anyone can sell it without paying you royalties.

Do I File with the U.S. Government?

Yes, you file with the PTO, just like you did for a trademark.

When Should I File?

Before you file, you should do several things to preprocess your application, which will save you time and money.

1. Prepare a disclosure of your invention, which includes a sketch and written description. If this sounds like your Inventor's Journal, you're catching on fast. The PTO has a Disclosure Document Program. For a $10 fee, you can list your invention in this program for 2 years. *This does not protect you like a patent does.* It merely shows you were there first.

2. Obtain a Record of Invention Form from a patent attorney. The cost for this form should be nominal, especially if you tell the attorney he or she may get your business for the patent search and subsequent application process.

3. Conduct your own patent search on possible competitors, using tools available to you. You can write the PTO and ask for information on how to use their search libraries and you can obtain a list of their libraries around the country. Several large universities also carry very extensive holdings of patents. Contact the school nearest you to find out.

NOTE: You cannot file a patent application if you have sold the product more than one year before filing the application.

What If My Patent Application Is Rejected?

Your patent application may be rejected for several reasons. Your claims may be too broad and unsubstantiated. They may be too obvious to a person skilled in the field of your invention. Or there may be earlier patents or publications that show features you are claiming as unique.

How Do I File for a Patent?

You can either file yourself, the hard way, or hire a registered patent attorney. Note that I said *registered.* Not just any attorney can represent you before the PTO. He or she must be registered with them, and has to have taken a rigorous exam. (For a listing of registered patent attorneys, check out the *Directory of Registered Patent Attorneys and Agents,* available from the Superintendent of Documents, Washington, DC.)

Patent agents are also registered with the PTO and can help in your application processing. However, they are not attorneys, and they are not qualified or authorized to defend your claims before the PTO or an appeal board.

There are very specific rules governing how your application should look and the order of the material presented. The order of arrangement, or material, is as follows:

1. Title of invention or preamble.
2. Other application under consideration.
3. Summary of the invention.
4. Description of the drawings.
5. Detailed description of form and function.
6. Claims.
7. Abstract.

Because these are very complicated, see your patent attorney.

The PTO Guidelines

The PTO normally designates five qualifiers as benchmarks for the granting of a patent. Your invention must be:

1. Original and practical.
2. Unknown.
3. Useful.
4. Workable.
5. Past the idea stage.

Original and Practical. If your invention is not original, what is the value to future consumers? None! I'm not saying it must be 100% original. But it must have some value that is totally original, in order to make it worthwhile for people to buy. After all, you want people to buy your invention, right?

Practicality is obviously important. An invention that is too complicated, and that has little practicality, will not sell either.

Unknown. The PTO looks for inventions that are unknown in the method you submit them in. They obviously cannot patent the same thing twice. If you change your invention just enough to make it different from and better than another's, you might get your patent. Check with your patent attorney, whose job is to show the examiner the differences. It's best to come up with something new and unique.

Useful. The third qualifier for granting a patent is usefulness. The PTO wants to see that your invention has a useful purpose in society, that it is not just some fad with no intrinsic value. For example, while the Pet Rock sold thousands in its heyday, you couldn't get a patent on it—no intrinsic value.

Workable. Your invention must work. This may sound ridiculous, yet every year, inventors try to get patents on inventions that look great on the drawing board, but will not function.

Past the Idea Stage. The PTO wants to see sketches and diagrams of a working prototype, not just a good idea. *You cannot patent an idea.*

What Happens Next?

After you file your application, one of two things will happen. In very rare instances, you could receive a Notice of Allowance (a letter telling you you've been immediately granted a patent). Once you pay the issue fee, you will be issued letters patent. But this is a very rare occurrence.

Most likely, you will receive an *action*. This is the first notice you get from the examiner, telling you the objections to your claims. Here is where your attorney earns his or her fee. The attorney replies on your behalf in an *amendment,* in which he explains and perhaps modifies the claims portion with your input, of course.

You may go around a few times. The examiner may still reject the claim because it still doesn't fulfill one or more of the requirements in the PTO guidelines (described above), or you may get your patent. If you're still rejected, you have the right to appeal. If your appeal is rejected, you lose, meaning no patent, and you can't reapply for that invention. (Try another idea.)

What About Search Firms?

While most patent attorneys employ search firms to do the legwork of checking competing patents, it would not be a good idea for you to hire one. Unless you have the time to check out half a dozen and subject them to intense scrutiny, I suggest you let your lawyer handle it.

What Will All This Cost?

Compared to your attorney's fees, the application fee will be modest. When asking your attorney for a quote, make sure the estimate is itemized: how much for the search, the original fee, the amendments, etc. You also should expect at no charge an initial consultation with any attorney about the cost of the whole process. Patents run from $2000 on up. The more times your attorney has to defend your claims against the PTO patent examiner's objections, the more it will cost.

Does the Patent Process Mirror a Trademark Search and Application Process?

Other than the fact that both federal offices are located in the same building, there are not any substantial similarities. It is really not very difficult to prepare an application to register a trademark, whereas preparing a patent application, and obtaining the proper wording for claims to provide meaningful protection, require a great deal of study and experience. I believe it would be foolish for anyone to attempt the entire process alone.

What Is a Patent Assignment?

You see these quite frequently in business publications. An employee has developed a new invention or product, and the patent is assigned to his employer. A patent assignment states that *All rights to the patent have been sold to the employer.* If the employee has been drawing a salary while working on the invention or idea, this practice can be dangerous. If you're in this situation, contact an attorney *before* you get ripped off.

What Is Patent Licensing?

Patent licensing gives someone the right to operate under a patent and to pay royalties to the inventor. The holder of the patent retains legal title to the invention; the licensee can make or sell the product.

Where Can I Get More Information?

Contact your local Government Printing Office, and ask for the booklet *General Information Concerning Patents,* #003-004-00619-6.

COPYRIGHTS

What Is a Copyright?

A copyright is the exclusive right of an author to publish, reproduce, or sell, for a specific number of years, his or her work.

Do I Need to Register My Copyright?

Like the trademark, you can use the copyright symbol, ©, to establish copyright on your work. But legally you have nothing to show to prove your copyright, unless you file an application. At the modest sum of $10, anyone who doesn't file is foolish.

What Are the Different Classes of Copyright Protection?

The classes that apply to inventors are:

1. *Class TX.* For published or unpublished nondramatic literary works, excluding periodicals or serial issues. This class includes a wide variety of works: fiction, nonfiction, poetry, textbooks, reference works, directories, catalogs, ad copy, compilations, and computer programs.
2. *Class VA.* For published or unpublished works in the visual arts. Examples are pictorial, graphic, or sculptural works, including two-dimensional and three-dimensional works of fine, graphic, and applied arts; photographs; prints; reproductions; maps; globes; charts; technical drawings; diagrams; and models.
3. *Class RE.* For renewal registration. For copyrights already in their first term of copyright on January 1, 1978. This class works for all renewals in this time period, regardless of the class of the original copyright.
4. *Class CA.* For supplementary registration. This class is for correcting errors in your original copyright form.

How Long Will It Take?

The Copyright Office does not acknowledge receipt of your application. They get over 500,000 per year. However, within 16 weeks from submission, you'll get either a certificate of registration, a letter or phone call asking for more information, or a rejection letter.

To be sure they received your application, it's a good idea to send your application, along with your fee and samples of materials to be copyrighted, via registered or certified mail. A copyright application form is shown in Figure 5.5.

When Does the Registration Take Effect?

A copyright registration is effective on the date that all the required elements (sample materials, fee, and application) are received in acceptable form.

FORM TX
UNITED STATES COPYRIGHT OFFICE

REGISTRATION NUMBER

TX TXU

EFFECTIVE DATE OF REGISTRATION

Month Day Year

DO NOT WRITE ABOVE THIS LINE. IF YOU NEED MORE SPACE, USE A SEPARATE CONTINUATION SHEET.

1

TITLE OF THIS WORK ▼

PREVIOUS OR ALTERNATIVE TITLES ▼

PUBLICATION AS A CONTRIBUTION If this work was published as a contribution to a periodical, serial, or collection, give information about the collective work in which the contribution appeared. **Title of Collective Work ▼**

If published in a periodical or serial give: **Volume ▼** **Number ▼** **Issue Date ▼** **On Pages ▼**

2

a **NAME OF AUTHOR ▼** **DATES OF BIRTH AND DEATH** Year Born ▼ Year Died ▼

Was this contribution to the work a "work made for hire"? ☐ Yes ☐ No

AUTHOR'S NATIONALITY OR DOMICILE Name of Country
OR { Citizen of ▶ ______ / Domiciled in ▶ ______

WAS THIS AUTHOR'S CONTRIBUTION TO THE WORK
Anonymous? ☐ Yes ☐ No
Pseudonymous? ☐ Yes ☐ No
If the answer to either of these questions is "Yes," see detailed instructions.

NATURE OF AUTHORSHIP Briefly describe nature of the material created by this author in which copyright is claimed. ▼

b **NAME OF AUTHOR ▼** **DATES OF BIRTH AND DEATH** Year Born ▼ Year Died ▼

Was this contribution to the work a "work made for hire"? ☐ Yes ☐ No

AUTHOR'S NATIONALITY OR DOMICILE Name of Country
OR { Citizen of ▶ ______ / Domiciled in ▶ ______

WAS THIS AUTHOR'S CONTRIBUTION TO THE WORK
Anonymous? ☐ Yes ☐ No
Pseudonymous? ☐ Yes ☐ No
If the answer to either of these questions is "Yes," see detailed instructions.

NATURE OF AUTHORSHIP Briefly describe nature of the material created by this author in which copyright is claimed. ▼

c **NAME OF AUTHOR ▼** **DATES OF BIRTH AND DEATH** Year Born ▼ Year Died ▼

Was this contribution to the work a "work made for hire"? ☐ Yes ☐ No

AUTHOR'S NATIONALITY OR DOMICILE Name of Country
OR { Citizen of ▶ ______ / Domiciled in ▶ ______

WAS THIS AUTHOR'S CONTRIBUTION TO THE WORK
Anonymous? ☐ Yes ☐ No
Pseudonymous? ☐ Yes ☐ No
If the answer to either of these questions is "Yes," see detailed instructions.

NATURE OF AUTHORSHIP Briefly describe nature of the material created by this author in which copyright is claimed. ▼

NOTE

Under the law, the "author" of a "work made for hire" is generally the employer, not the employee (see instructions). For any part of this work that was "made for hire" check "Yes" in the space provided, give the employer (or other person for whom the work was prepared) as "Author" of that part, and leave the space for dates of birth and death blank.

3

YEAR IN WHICH CREATION OF THIS WORK WAS COMPLETED **This information must be given in all cases.** ______ ◀ Year

DATE AND NATION OF FIRST PUBLICATION OF THIS PARTICULAR WORK **Complete this information ONLY if this work has been published.** Month ▶ ______ Day ▶ ______ Year ▶ ______ ◀ Nation

4

COPYRIGHT CLAIMANT(S) Name and address must be given even if the claimant is the same as the author given in space 2.▼

See instructions before completing this space

TRANSFER If the claimant(s) named here in space 4 are different from the author(s) named in space 2, give a brief statement of how the claimant(s) obtained ownership of the copyright.▼

DO NOT WRITE HERE / OFFICE USE ONLY

APPLICATION RECEIVED

ONE DEPOSIT RECEIVED

TWO DEPOSITS RECEIVED

REMITTANCE NUMBER AND DATE

MORE ON BACK ▶ • Complete all applicable spaces (numbers 5-11) on the reverse side of this page. • See detailed instructions. • Sign the form at line 10.

DO NOT WRITE HERE

Page 1 of ______ pages

Source: U.S. Copyright Office, 1987.

FIGURE 5.5 Copyright Application Form

How Long Does a Copyright Last?

Under the new law, a copyright is good for the life of the author, plus 50 years.

Where Can I Get Help?

For further information on copyright classes, forms, fees, etc., contact the Register of Copyrights, Copyright Office, Library of Congress, Washington, DC 20559. Or call the Copyright Office Hotline at (202) 287-9100 and leave a message on their recorder.

PART III

THE MONEY GAME

6

Financing Your Invention: From Seed Money to Expansion

You're at the point now where you need to get seed money to make a prototype. Or perhaps you've got the model done, have a few feelers out, and are beginning to plan for production. This takes money, too.

So, where do you come up with the money? Most people try to tap their friends and relatives. If you're lucky enough to have family with vision, great! For the rest of us, finding capital to fund our inventions and projects is a very difficult job. I hope to make it easier for you by giving you the tools you need.

WHAT YOU WILL NEED

For any type of funding you intend to pursue, you must understand two basic facts about potential lenders:

1. All lenders will want to see that you have a working knowledge of running a business and know where you want to go with your product. If you don't have all the expertise (and no one does), you'll need to have partners or outside experts ready to help you. You must also understand basic marketing and know who your targets are. Most funding sources are very leery of inventors who do not know their target markets and where these customers can be found.

2. Every lender has one important goal in mind when lending money: How soon can he get his money back at a profit? If you understand this, you will be able to negotiate more effectively.

If you hope to sell anyone on funding you, you have to sell yourself and your invention. Here's what you'll need for a presentation to a lender. It's always best to be overloaded with material, and not need it, rather than be short of information, and have to fudge.

1. A personal financial statement.
2. A business plan with all the elements.
3. A prototype.
4. A statement of the amount of money you need, and what you need it for.
5. A profit and loss statement.
6. A prospectus.

The first four items are discussed in various chapters in this book, so there's no need to repeat them here. The profit and loss statement is one that your accountant or bookkeeper can furnish you. (You will not have one when you start. Be sure to have a projected statement of sales and expenses for 1 year, a cash-flow forecast, and a current balance sheet. See Chapter 3 for details.)

A prospectus is used when raising money through limited partnerships, private placements, and stock offerings. As this is a very sophisticated way to raise money, contact your attorney and a competent stockbroker or financial advisor to get help.

THE BUSINESS PLAN

As you read in the previous chapters, every good business starts with a plan, and yours is no exception. Whether you intend to seek venture capital or go for a grant or loan, you need a sound plan.

The outline of a good business plan was discussed in Chapter 3. Now would be a good time to reread that section. To review briefly, a good plan *must* include the following:

Description of the business
Company history
Target markets
Competition
Management and personnel
Marketing plan
Distribution plan
Financing plan
Production information
Financial data
Any supporting documents

FUNDING SOURCES

There are many ways to finance your invention's development and production. It takes guts, determination, and a willingness to "do-it-by-the-numbers," to wade through the red tape.

Each program, whether grant (free money) or loan (you must pay it back), has certain restrictions on it. These may be related to geographic location, industrial classification, or ethnic background. Check with the lender's guidelines first, to save time.

Below is a list of possible sources to consider. Use your imagination, and you'll see other options as well.

Accountants
Banks; Savings and Loans
Business development corporations
Credit unions
Family and friends
Foundations and grant programs
Incubators

Inventors' clubs
Insurance policies
Manufacturers and corporations
Personal funds
Private placement of stock
Real estate equity loans
State and local government programs
U.S. government programs
Small Business Innovation Research Programs (SBIR)
Small Business Investment Companies (SBIC)
Venture capital clubs
Venture capital companies

How many should you investigate? As many as you can, but only as many as you can do well. Don't put all your eggs in one basket.

Accountants

While most accountants are not in the money-lending business (although if your accountant is a true member of your team, he or she may see the wisdom of investing in your project), they know where money is. Ask yours if he or she has other clients with money to invest.

Banks; Savings and Loans

These are probably the hardest to tap during the seed phase. (Especially with the current scandals, it's almost impossible.) They want to see a proven track record of at least 2 years in business. And how do you get 2 years' experience without a loan? It's the Catch-22 syndrome.

Unless you use a savings account as collateral (in which case, you're borrowing your own money, something I abhor), or have good enough credit for a signature loan, don't count on your friendly neighborhood bank. But do try a local bank that knows you before trying a large chain.

1. For listings of banks around the country, write the Independent Bankers Association, 1168 S. Main St., Sauk Center, MN 56378.

2. The SBA has a special lenders' program called the PLP, or Preferred Lenders Program. This program is made up of banks that give a hefty percentage of SBA loans, which makes them more favorable to you. They are licensed by the SBA and follow strict rules. Call the SBA Answerdesk at (800) 365-5855 for a complete list.
3. *Polks Bank Directory* also lists banks around the country. Check the reference section in your county library.

Business Development Corporations

Be suspicious of business development corporations that promise you the moon and all the financing you'll need "if your product is marketable." While there are some who do what they say they will, these are rare. Check them out thoroughly with your lawyer or accountant.

Credit Unions

A credit union is often an overlooked method of financing the start-up of a small business. Some credit unions will even offer you a signature loan up to $12,000 with good credit. Of course, you must be a member. Some credit unions may also have an inventors' program, where you can get matching funds if your employer is involved in the project (like the aerospace industry). Check it out.

Family and Friends

Although family and friends should be your starting place for funds, most of them tend to pooh-pooh your idea. Why should you subject yourself to criticism and ridicule? Because it thickens your skin for when you have to negotiate with the real lenders. Listen carefully to their criticisms; they may have valid points other lenders will also ask.

Foundations and Grant Programs

These can be excellent sources of funding for the inventor. As with all lenders, you must have your act together. (Reread the beginning of this chapter about what you'll need.) The following directories and books list foundation and grant programs. If you've never written a grant proposal, I suggest you get a good book on the subject, or hire a freelance writer specializing in grant proposals. (See *Writer's Yellow Pages* for such a list.)

Foundation Center Directory (yearly book)
312 Sutter Street #312
San Francisco, CA 94108
(800) 424-9836

This book lists about 2000 foundations and their requirements. It is usually in the county library reference section.

Grantsmanship Center
650 S. Spring Street #507
Los Angeles, CA 90014
(213) 689-9222

This center presents seminars on how to get grants.

Foundation Directory
Gale Research Company
Book Tower
Detroit, MI 48226
(800) 223-GALE

Gale Research publishes various grant and foundation directories.

Free Money for Small Business and Entrepreneurs
by Laurie Blum
John Wiley and Sons
605 3rd Avenue
New York, NY 10158

This is an excellent book loaded with names and addresses of foundations and grant programs, listing deadlines, requirements, etc.

Incubators

Incubators are run by nonprofit groups to help start-ups. They often provide inexpensive office space, clerical assistance, computer and office machine services, and advice for small businesses. I suggest you subscribe to *Incubator Times,* an SBA quarterly magazine, which lists various incubators in the United States. For a complete listing, write to the National Business Incubator Association, 114 N. Hanover Street, Carlisle, PA 17013.

Inventors' Clubs

While inventors' clubs per se are not sources of funding, they can lead to money sources through their networking ability. (See Chapter 15 for a list of clubs.)

Insurance Policies

Some insurance policies let you borrow against your accrued cash value. But be careful. If your invention doesn't sell, you still have to pay back your policy. And, of course, you are charged interest.

Manufacturers and Corporations

If your invention is exciting enough, you may be able to sell out to a manufacturer or corporation. In this case, you don't need seed money. The invention is no longer yours. But you might be able to get a manufacturer who is going to produce your invention to finance you or give you credit during the production phase. This can be a tough road, unless you are a good negotiator. (This is the route most commonly hoped for by inventors. It's also the most frustrating. Making a 10% royalty on a dollar is not my idea of time well spent. And how will the manufacturer guarantee sales?)

Personal Funds

Anytime you invest your own funds, you show other lenders the faith you have in your own invention. Just be sure not to play your entire hand at once. Personal funds include savings accounts, real estate equity, insurance cash value, etc.

Private Placement of Stock

This is like a glorified personal loan pool from friends, families, and interested parties. You give each "investor" a share of stock in your company, in exchange for a set amount per share. (Be sure you're incorporated, and follow the SEC rules.) This is the way I financed a third round of funding for a magazine I owned in San Diego several years ago. Make sure your investors understand the risk involved.

Real Estate Equity Loans

If you own real estate with decent equity, this is an easy way to obtain business capital. A second or third trust deed on your home, or equity financing, can give you roughly 70% of the equity value. But don't bet on the entire amount of equity in your home or property.

State and Local Government Programs

There are many state and local government programs for grants and small business loans. Contact your local city hall, or state department of commerce, for a list of available programs and requirements.

    An excellent source of program listings is Matthew Lesko's book, *Government Giveaways for Entrepreneurs,* Information USA, P.O. Box 15700, Chevy Chase, MD 20815, (301) 657-1200. I refer to it all the time. (If you talk to Matthew, tell him I sent you.)

U.S. Government Programs

There are literally hundreds of government programs giving away money every year. As an inventor, your job is to track them down and write a proposal for *each one* that fits your company or invention. Some sources for these programs are:

1. Matthew Lesko's book. See above.
2. *Catalog of Federal Domestic Assistance.* Write to the U.S. Library of Congress, Washington, DC.
3. The Small Business Administration should be a starting point for loan assistance. A few of the loans they have are:
   - Loans for Low-Income Entrepreneurs, SBA 59.003.
   - Loans for Businesses Turned Down Elsewhere, SBA 59.012.
   - Loans Through Local Governments, SBA 59.013.
   - Energy Loans for Inventors, SBA 59.030.
4. The U.S. Department of Agriculture offers two loans that may help:
   - Small Business Innovation Research (SBIR), USDA 10.212.
   - Businesses/Towns Less Than 50,000 People, USDA 10.422.

Small Business Innovation Research Programs

The purpose of this program (SBA 13.126) is to stimulate the technology of government-funded research. Contact your local SBA office. They will help you with document preparation. This program is funded by:

Office of Innovation Research and Technology
Small Business Administration
1441 L Street, NW, Room 500-A
Washington, DC 20416
(202) 653-6458

Another program to help inventors is from the National Bureau of Standards. Their program (81.036) gives grant money to inventors of energy-related projects. Write them at National Bureau of Standards, Gaithersburg, MD, or call (301) 975-2000.

Small Business Investment Companies (SBIC)

There are over 140 SBICs licensed by the Small Business Administration. They are venture capitalists, regulated by the SBA. For a complete list, write the SBA in Washington, DC, or call your local SBA office.

Venture Capital Clubs

These are like inventors' clubs. While they do not have funds to disperse, nor do they raise capital, they are excellent networking groups that can help with finding sources and preparing documents. For a complete listing of clubs, write to:

> Venture Capital Network
> c/o Dr. W. Wetzel
> P.O. Box 882
> Durham, NH 03824

Venture Capital Companies

Most venture capital companies do not work with start-ups. You really have to have your act together. For a listing of sources, see Stanley Pratt's book, *Guide to Venture Capital Sources,* in the library.

MONEY FINDERS

Money finders or brokers *can* be helpful in your search for funds, provided you deal with a reputable company. Like invention brokers, the legitimate ones do not charge exorbitant finder's fees for their services. The best starting place would be inventors' clubs or venture capital clubs. Ask the members who they know.

UNDERSTANDING THE TERMINOLOGY

It is necessary to be *totally* prepared when approaching any funding source. And it helps if you know the lingo. I suggest you understand the basic bookkeeping terms like assets, liabilities, debits, credits, etc. It also helps to know specific venture capital and funding terms. Check your library for books about funding.

An excellent new book is Gene Valdez' *How to Prepare a Bank Financing Proposal for your Business.* You can get it by calling (909) 621-6336.

A FEW FINAL SOURCES

Assuming that you have what it takes (and who wouldn't take the time, with all that lovely free money just waiting for you), here are a few more of the many programs available for inventors.

> NBS Office of Energy-Related Inventions
> National Bureau of Standards
> Washington, DC 20234
> (202) 921-3694

This program encourages you to develop non-nuclear energy technology. They help individual and R&D companies working on energy-related inventions. Write for an application and details.

> Office of Planning and Coordination
> 14th Street and Constitution Avenue, NW
> Room 3223
> Washington, DC 20230
> (202) 377-4921

This office encourages local businesses to enter into and expand their export marketing programs.

> The Xerox Foundation
> P.O. Box 1600
> Stamford, CT
> (203) 329-8700

This foundation helps companies that advance knowledge in science and technology.

> Information Officer
> Overseas Private Investment Corp.
> Washington, DC 20227
> (202) 653-2800

This organization helps companies and individuals with private overseas investments. This is a sophisticated program and should be approached with a consultant.

I've saved the best for last:

Mancuso's Small Business Resource Guide
by Joseph R. Mancuso
Prentice Hall Press, 1988

This is another *must* for any inventor's library. It lists hundreds of funding and entrepreneurial assistance sources.

7

Budgeting and Scheduling

WHY BUDGET AND PLAN?

The inventor who does not plan has a fool for a partner. The inventor who does not set a budget is a poor fool.

This chapter will outline the various budgeting methods for marketing: the old ways *and* the new way. While the earlier methods did their job, that was then, and this is now. Modern competition *demands* that a marketing budget be precise, lean, and on target! It is precisely this "on-target approach" that we will explore.

We will also discuss scheduling of marketing events. Each invention you create will become part of your overall company marketing plan. These segments, or campaigns, must be well thought out, both strategically and tactically. If this sounds like you're going into battle, you are!

The average human brain receives over 72,000 sensory impressions a minute. During the day, we receive close to 16,000 media impressions. *The competition for our attention is tremendous!*

You must, therefore, plan your marketing like you were going to war. In fact, one of the all-time greatest marketing books, *Positioning: The Battle for Your Mind,* by Al Ries and Jack Trout, is dedicated to Carl Von Clausewitz!

Von Clausewitz, if I remember correctly from my college political science classes, was one of the world's leading military strategists.

And you were wondering why marketing buzzwords sounded so warlike? Campaigns, strategies, tactics, guerrilla marketing—makes sense now, doesn't it?

BUDGETING: SETTING THE RECORD STRAIGHT

To carry out a proper marketing budget, you need to keep good business records from day 1! Record *all* your expenses and income, every day. Make sure you follow the suggested small business guidelines for general record-keeping and budgeting (see the examples below).

Decide immediately to set aside part of each day to update your business records. Assume that something might happen to you within the next 8 hours; if it should, your records should be in such order that anyone could pick up the threads of the business with minimum knowledge of where you stand.

TYPES OF RECORDS

According to the SBA, in the most basic sense, business records should tell you three things:

1. How much cash you owe.
2. How much cash is owed you.
3. Cash you have on hand.

Your checkbook is not the only set of records you keep. The checkbook balance is not necessarily a true picture of your business.

You must have five basic journals for your business. It is not necessary for you to constantly keep them updated yourself, but they must be available. It's a good idea for you to understand how they operate. (Hire a bookkeeper, and make sure you know enough so you can make sense of things if the bookkeeper leaves.) You might also get a book on bookkeeping skills, or take an SBA or junior college course.

- *Check register.* Shows each check disbursed, the date of disbursement, the number of the check, to whom it was made out (payee), the amount of money disbursed, and for what purpose.
- *Cash receipts.* Show the amount of money received, from whom, and for what.
- *Sales journal.* Shows the business transaction, date, for whom it was performed, the amount of the invoice, and the sales tax, if applicable. It may be divided to show labor and goods. (A sales journal is very important to you as an inventor.)
- *Voucher register.* Records bills, money owed, the dates of bills, to whom owed, the amounts, and the product.
- *General journal.* A means of adjusting some entries in the other four journals.

CASH FLOW

If you begin with $10,000 cash at the start of the year and have $300 at the end, you'd definitely want to know what happened! This does not necessarily mean you had a bad year. You might find the cash went for machining, tooling, inventory, etc. You have to find out.

Cash flow is tracking the movement of business funds through the business cycle to see where it goes. In Chapter 3, you did an estimate of cash flow. Go back and reread this section, and make sure you understand it.

BREAK-EVEN ANALYSIS

Part of your initial business plan should be to estimate your break-even point—a point in business where you are neither making money nor losing money. It shows how much you must sell to cover your costs.

In an invention or manufacturing business, you should determine at what approximate level of sales a new product will pay for itself and begin to bring in profit.

Profit, of course, depends on sales volume, selling price, and costs. To figure your break-even point, separate your fixed costs, such as rent, from your variable costs per unit, such as labor and materials.

The formula for break-even volume is:

Total fixed ÷ Selling price per unit minus
Cost   Variable cost per unit

As an example, if you determine your fixed cost to be $200,000 per year, your variable cost to be $40 per unit, and selling price to be $60 per unit, then your break-even volume (BEV) is 10,000 units:

BEV: $200,000 ÷ $60 − $40 = 10,000 units

Of course, in order to lower your break-even point, you can lower your fixed costs or your variable costs. If you have determined a maximum selling price, with a small margin, you've got to move units to make it!

SETTING YOUR SALES PRICE

To make it worth your while, your invention should sell for four or five times your total variable cost per unit. If your product sells for $20 and you sell it through normal distribution channels (see Chapter 9), you will only get about $6 for it. And that's before you deduct the actual cost of making it!

To see how this works, let's take a $20 product:

Selling price	$20.00
50% and 10% discount	−$11.00 (wholesale)
Wholesalers cost	$ 9.00
Freight, commissions, discounts	−$ 1.44 (4%, 10%, 2%)
Net to you before your expenses	$ 7.56
Overhead costs (not inc. mfg.)	−$ 1.51
What you realize for your product	$ 6.05

Cost per unit (materials, labor)	– $2.50
Net Profit to You Per Unit	**$3.55**

Scary, isn't it?! You bet! Most inventors don't realize what it takes to market and sell a product. Let's break down each of the above items, so you understand them clearly.

1. Selling price is four to five times (sometimes more) your manufacturing cost per unit. The manufacturing cost includes the finished product *plus* all packaging costs.

2. The discounts are industry standard for wholesalers or jobbers. Why? Because they sell to retailers, and retailers want at least a 30% markup, sometimes even 40%. Note that the discounts are on different values. The first discount comes off the full selling price, and the second comes off the remaining balance. (This is just the way it is. Mail-order houses and catalogs are quite demanding. You have to play by their rules.)

3. Freight costs usually run about 4%, sales commissions to manufacturer's reps about 10%, additional miscellaneous discounts another 2%. It adds up, doesn't it?

4. Now you have a net figure *before* any expenses. Next, take off overhead for your office, supplies, heat, car maintenance, advertising, etc., prorated on a normal 20% basis. What you have left is what you get for your invention, *before you figure how much each unit costs, with its packaging.*

5. Finally, subtract your unit cost, and what you have left, if anything, is profit.

This depresses even *me!* What we need are fewer discounts to wholesalers (probably not in my lifetime), or lower manufacturing costs (why many go offshore), or buyers who realize they have to pay a decent price for products (my hat's off to those of you out there).

SETTING YOUR MARKETING BUDGET

There are five methods of budgeting. Four of them are old hat. One is the only method that truly makes sense in today's marketplace, given the amount of competition. Below are the five tried and trues.

Percentage of Sales

This is probably the oldest and still most commonly used method. Many marketing and advertising texts tell you to set aside 5% of your gross annual sales for promotion and marketing. Then they say, if you're just starting, or introducing a new product, or you have major competition, you need to think 10–12%.

The problem with this method is that it doesn't account for any seasonal fluctuations, nor does it consider what the competition is doing.

Match the Competition

This method says, "Whatever he spends, I'll spend!" *Really dumb!* What if he has more money than you? What if his fixed or variable costs are different? (They probably are.) What if this is a ploy to get you to blow your cash real fast, and then he backs off, while you suffer? How do you know he didn't barter product for promotion? How do you know what rate he got for his advertising? *You don't!*

All You Can Afford

This is another nonsensical way to waste your marketing dollars. How do you know really what you can afford? You won't know until the end of the campaign, to see if it works. Then you'll know if you could have afforded it. Guess what? Then it's too late!

Last Year + X%

This method says to add X% to last year's budget. This doesn't work, either. Costs go up every year. This method doesn't consider inflation, the cost of goods sold, or anything else. It's not really a viable method.

The Objective-and-Task Method

This is the only logical method to use. First figure out what your marketing campaign objectives are for a given period—say 1 year. Then determine what tactics (marketing methods) to use to meet these objectives. List them. Then find out what each task will cost. The amount of money it takes to accomplish the tactics determines the budget, not vice versa.

A SIMPLE METHOD FOR RECORDING A BUDGET

To get a handle on the objective-and-task method of budgeting, you need to do the following:

1. Learn all you can about each method's costs (from media kits, market research, and other inventors and entrepreneurs).

2. Using the sample monthly budget guideline chart in Figure 7.1, try to determine what a possible marketing mix would be for a hypothetical invention. (You need to know the target audience, and what they respond to—radio, billboards, newspapers, etc. For more information on target audiences, see Chapter 11.)

3. Fill in the sample monthly budget guide. Go for broke! Make a wish list of all possible methods you would use. If it seems too steep, cut out some of the methods.

4. When you complete your sample budget, based on the tactics you intend to use, you will have determined how much you need to spend.

THE SCHEDULING TECHNIQUE

Plan everything you possibly can. The first thing you'll need, a calendar, has a threefold purpose:

1. Other events may conflict with your scheduled event or campaign.

2. Other events may prove helpful to you; you might piggyback on them to your advantage.

3. Your campaign will have to be scheduled well in advance, and it will cover several months.

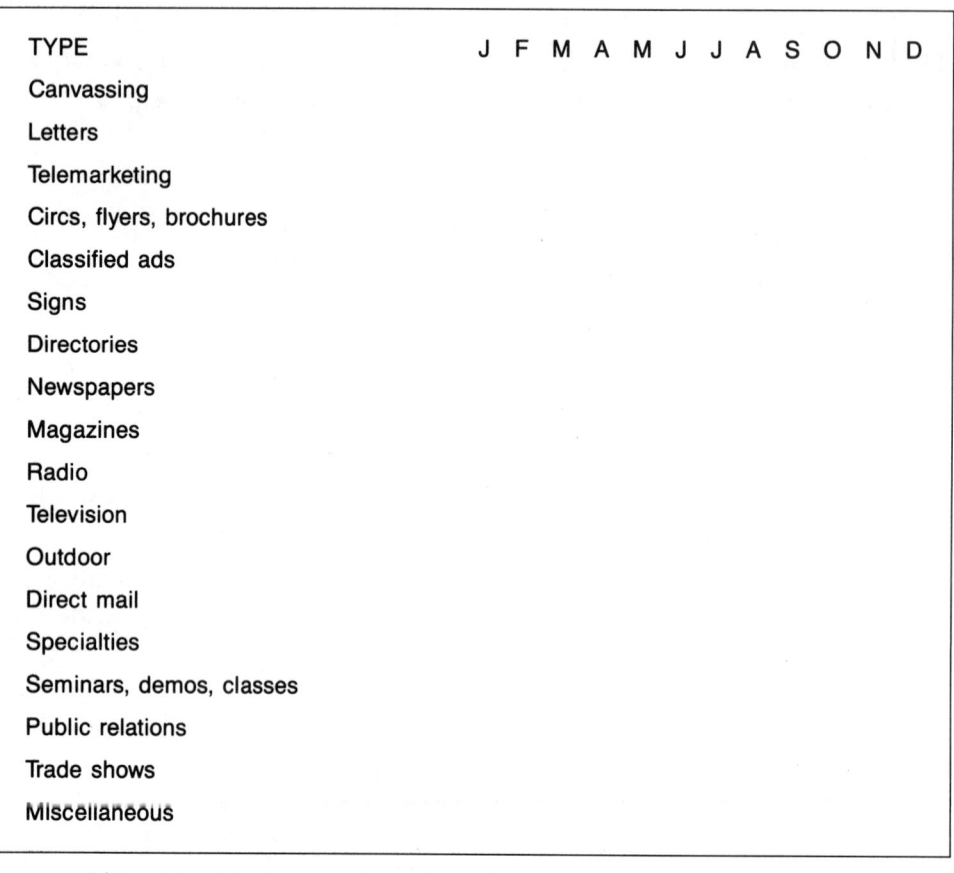

TYPE	J F M A M J J A S O N D
Canvassing	
Letters	
Telemarketing	
Circs, flyers, brochures	
Classified ads	
Signs	
Directories	
Newspapers	
Magazines	
Radio	
Television	
Outdoor	
Direct mail	
Specialties	
Seminars, demos, classes	
Public relations	
Trade shows	
Miscellaneous	

FIGURE 7.1 Monthly Budget Guideline Chart

Here's how to schedule your marketing:

1. Get a yearly calendar with month-by-month pages.

2. Block out all holidays.

3. Write down every seasonal event (for example, the Super Bowl, World Series, L.A. Marathon, local happenings).

4. Check library reference sections for an almanac of special events, weeks, or months that you can piggyback on. Pencil these in. (Try Chase's *Annual Directory of Events*).

5. Check with your local chamber of commerce for their events, and pencil them in.

6. Call local newspapers. Ask the calendar editor for upcoming events during the next 6 months.

NOVEMBER 1989

SUNDAY	MONDAY	TUESDAY	WEDNESDAY	THURSDAY	FRIDAY	SATURDAY
			L.A. Times Ads - 1 Day Sale 1	Direct Mail Piece For 1-Day Sale - to Post Office 2	3	1 Day Sale 9-9 4
5	Promotion Meeting	Pre-Holiday Private Sale Mailer Out 7	L.A. Times 1/2 Pg. Ad KNX/KFWB 8	L.A. Times 1/2 Pg. Ad 9	*Competitor SALE Check it Out! 10	SALE 9-6 Pre 1-Day sale 11
Scout Sunday Sponsor 12	13	Final Proofs needed - New Years Ads 14	Speech to Lion's Club 15	Pre-Holiday Private Sale to Special Christmas 16	17	Cooking Demo Housewares 18
19	20	Mail New Year Ads to all Media 21	Radio/TV Blitz - Start X-mas sale 22	THANKSGIVING ✗ 23	SUPER SALE Pre-Christmas 9-Midnite 24	25
Sale 9-6 Pre X-mas 26	27	28	Budget Meeting 29	Planning Meeting 30		

FIGURE 7.2 Promotion Schedule

89

7. You'll end up with several slots or holes during the month where very few conflicts occur. These are ideal for holding ad campaigns, press events, and other marketing methods for your invention. The less competition you have, the greater the chance of getting press coverage.

8. Pick one hole per month, and pencil in some kind of marketing event, an ad, press release, or brochure, for example. Although tentative, this will get you visually considering how much time you need for implementation.

Once you have penciled in a project every month, next determine which tools to use to achieve visibility for your invention. See Figure 7.2, on the previous page, for an example.

PART IV

SELLING YOUR INVENTION

8

Selling Out for Big Profits: Finding a Buyer

SHOULD YOU SELL OUT?

The dream of every entrepreneur is to build a business to the point of selling out at a maximum profit. At least, it's the dream according to most small business magazines. But is it yours?

The problem is, it may be just that—a dream. You may end up getting far less than you hoped for your business, should you find a buyer. Then again, you *might* sell to a major corporation. But I wouldn't plan on it yet.

The purpose of this book is to show you how to market your own inventions and make money in the process. I would be remiss, however, if I did not include a chapter to help those of you who have decided the time is ripe to find a buyer. So hang on, here goes!

SELLING OR MANUFACTURING: THE BIG DECISION

Either road you take, there's more than meets the eye. Let's take a look at what's involved in manufacturing your invention first.

There are three ways to manufacture your invention: doing it yourself, subcontracting the labor and materials, or farming out the whole thing.

Doing It Yourself

You may decide to manufacture your product yourself. This is a big job, and I don't recommend it. Your manufacturing costs must be low enough for you to make a profit. The simpler the invention, the easier this can be. To find suppliers for the raw parts, tooling, and machines you'll need, check the business-to-business yellow pages, and other directories like the *Thomas Register of Manufacturers*.

Subcontracting

In subcontracting, you are the manufacturer, and you control what happens, while others get the materials for you and develop them into finished products. This is how Kiazi, Inc., works. This small jewelry company would not interest any sizable manufacturer just yet. The business is still in the embryonic stage. But they can't afford the time or money to do their own manufacturing.

Solution: subcontract for the acrylic jewelry. One person supplies the raw material, one silk-screens it, and one heats and bends it into shape. The last one then delivers a finished product. (They still must coordinate the entire process, however.)

Farming It Out

This is the best method of all three for the start-up inventor. Write down every step involved in the manufacturing of your product. For example, in Kiazi's case, there are five processes: buying the acrylic, cutting it, silk-screening, heating and bending to desired shape and length, and packaging.

In farming it out, the person you deal with will normally be the one who furnishes the raw goods. Let's continue with the Kiazi example. The person who buys the acrylic and cuts it, the first contact person, acts as the general contractor. His responsibility is to make arrangements with other vendors to supply the other necessary steps. He then turns over the packaged, finished products. Kiazi never has to be involved in each step.

A word of caution here. Find someone you can trust to be your general contractor. Get estimates from six different companies, and pick the best one.

One final note. How do you pay the general contractor for coordinating the entire process? Offer him an additional 10% to take charge of it. You'll find him eager to help.

SELLING YOUR INVENTION: THE FOUR-STEP PROCESS

If you decide to sell, there is a four-step process you must follow to maximize your investment in time and money.

1. *Find a company.* Use the following sources (available in the library) to find manufacturers and corporations able to buy your invention:

Thomas Register of Manufacturers
Standard and Poor's Register of Corporations
Dun & Bradstreet Million Dollar Directory
Dun & Bradstreet Middle Market Directory
Trade directories
Trade magazines

2. *Know the company.* Before dealing with any company, know who the players are. Find out about their location, economic health, reputation, and size. (Use the above directories.)

3. *Have a disclosure form.* A disclosure is a written description of what you are selling, its history, and what you want. It should include the following nineteen points:

    a. Descriptive name of invention.
    b. Purpose or use.
    c. What is new and advantageous about it.
    d. Drawings and descriptions of parts.

e. Explanation of construction and operation.
f. A U.S. patent.
g. Patented abroad?
h. If unpatented, has a favorable search been made?
i. Patents found in the search.
j. A model or artwork for inspection.
k. Who needs the invention most?
l. Have any manufacturing efforts been made?
m. Have any completed goods been made?
n. Are there any tools or dies available?
o. Have any been sold?
p. Were there any repeat orders?
q. What advertising has been done?
r. Do you want cash or royalty?
s. Will you help in developing the invention?

4. *Protect yourself.* With any disclosure, you have to protect yourself legally. Don't discuss or show your invention until you have a signed nondisclosure agreement from the parties you are dealing with. This agreement basically states that the invention is yours, and they will not discuss it with anyone else for 30 days. At the end of the 30-day period, if they don't want it, they return all material to you, and keep their mouths shut! (See Chapter 5.) Put the nondisclosure agreement in an envelope on the outside of the package you're sending. Include the following note on the envelope: *"Warning: This package contains proprietary materials. If not willing to sign a nondisclosure agreement, do not open package, and return intact."*

HOW TO INTEREST A PROSPECTIVE BUYER

There are four steps that will make it easier for you to interest potential buyers:

1. Use a model or prototype.
2. Use your own drawings and descriptions.
3. Use your patent application.
4. Use a disclosure form.

Model or Prototype

This shows the potential buyer (or investor, for that matter) that you are past the idea and drawing-board stage, that you are ready for his help, and hopefully, his money.

Drawings and Descriptions

The buyer will want to see that this is indeed *your* invention, not someone else's. He will also want to see how you write and draw. This tells how good an inventor you are from the standpoint of engineering, patents, etc.

Patent Application

Show the buyer your patent application, and he will know you're one step closer to making the product a reality. It's one less step he has to do, and one closer to making money.

Disclosure Agreement

Any legitimate buyer will respect an inventor who brings a nondisclosure agreement to sign.

OTHER WAYS TO FIND A BUYER

There are other ways to find buyers for your invention. You might consider running classified ads in business sections of newspapers, or using an invention broker.

Classified Ads

Run a classified ad in either the business opportunities, investment, or invention-for-sale section of your local newspaper. (I'm assuming your invention is protected, right?) Make sure you include these six parts of the ad, at minimum:

1. Name of product category (e.g., electrical invention).
2. Description of product.
3. What the product will do.
4. Market(s) for the product.
5. Terms of the sale.
6. Who to contact.

Write your ad as descriptively as possible, so the reader will be excited. Sell the sizzle, and you'll get them to eat the steak later! But don't make it too long, or they'll get bored. And it will cost too much money. The example below should help. (For further discussion of how to write classified ads, see Chapter 13.)

INVENTION FOR SALE!

Electrical Invention for Household Use. New Labor-Saving Device. MASS Market Appeal. Sale Terms Negotiable. Call Bright Ideas (909) 555-3241.

Using an Invention Broker: Caveat Emptor!

An invention broker is someone who helps invention buyers and sellers get together, and who makes a commission. They may have contacts with various manufacturers, who will look at your product and analyze its potential. My opinion, and that of authors of other books I've read, is to consider invention brokers very carefully.

As an experiment, I picked up two months' worth of *Popular Mechanics* and *Popular Science* and called a dozen so-called invention brokers at random. I asked them to send me information. Several tried to sell me on themselves right then. When I showed reluctance and asked for literature to check them out, a few made it very clear what they were after. (From one, $400 in advance, *before* he would even look at my product!)

"Well, Mr. Franklin," he said. "Are you interested in having us help you?" (The "or not" was implied.) "If you are, send us back the forms we'll send out, with a check for $400, and we'll get started right away."

Chapter 8 Selling Out for Big Profits: Finding a Buyer   99

The funny thing is, I hadn't even described my invention yet. (Of course, I'd never tell these guys about a real invention, or describe it on the phone.) I said I had an invention and needed help in patenting and marketing it. Or perhaps I might want to sell it—could they help? Of course, they could.

While there are some legitimate companies out there, I tend to shy away from those who promote themselves too much. My advice is, *stay away from invention brokers.*

WHERE CAN YOU GET HELP?

Remember, what you're looking for is an honest, low-cost evaluation of your invention. There are certain places that do qualify.

These places do not market inventions; this is a point in their favor—no prejudice. Also, they only give positive ratings to *relatively few inventions*—a second point in their favor. Most invention brokers would tell you the moon was a great product, in order to get your money.

Actually, these places are universities. Universities are not in it for profit, which is great. And they have a reputation to maintain. For nominal costs, you get scientific evaluations, done by professors, who feed the data into a computer scoring system.

There are over 300 universities in the United States with technology centers that will evaluate your ideas or products for less than a few hundred dollars. For a list, see Dick Levy's *Inventor's Desktop Companion* (Detroit: Visible Ink Press, 1991).

I received material from two randomly selected universities:

WISC (Wisconsin Innovation Service Center)
402 McCutchan Hall
University of Wisconsin-Whitewater
Whitewater, WI 53190
(414) 472-1365
$150 flat fee

Baylor University
Innovation Evaluation Center
Hankamer School of Business
Waco, TX 76798
(817) 755-2265
$100 flat fee

LICENSING AND SALE OF YOUR PATENT

There are two ways to sell your invention: Sell it outright and have nothing more to do with it, or license the rights to it, so you still have some control. Let's explore licensing, which I believe to be the wiser of the two choices.

The Twelve Questions

There are twelve questions that must be answered *before* you license your patent. The questions are:

1. Is the license to be exclusive or nonexclusive?
2. What is the term of the license?
3. How is it renewable?
4. How will the royalties be paid?
5. When do the royalties start?
6. How often are they paid?
7. What are the guaranteed minimum yearly payments?
8. What happens if the terms are not met?
9. How are you to be involved?
10. What if the licensee infringes on your patent?
11. How do you end the deal if you're not happy?
12. How much should your lawyer be involved?

Two Key Points

Most of the above questions are self-explanatory. The two key questions are about exclusive and nonexclusive rights (#1) and your attorney (#12).

An exclusive license grants a company the right to market and sell your invention exclusively within a certain territory. No one else can encroach on their area. A nonexclusive license permits others to sell and market your invention in the same territory as the first party, perhaps with the same guidelines.

If the license is to be nonexclusive, what are the limits to it? You must decide. (Personally, I don't think you should give anyone an exclusive on anything; you lose too much control.)

By now it should be obvious that your lawyer should be involved in *all* aspects of your invention, especially if you sell part or all of the rights.

9

Distribution: Moving It on Out!

Most marketing books talk about distribution in a theoretical sense. They use terms like channeling, product flow, and others that no one understands. Distribution in their sense of the word deals with product movement.

Distribution in this book means selling!

THE MANY WAYS TO SELL YOUR INVENTION

There are dozens of ways to sell your invention in the marketplace. But not all ways will be good for your product. Depending on what it is and what market(s) it serves, some methods may be totally wrong.

For example, I would not sell a computer invention through a party plan or door to door. I would sell it through direct mail, manufacturer's reps, wholesale distributors, and retailers.

THE COMMON DISTRIBUTION METHODS

The most common method of distributing (selling) your product is from you to a wholesaler to a retailer to the consumer. There is nothing wrong with this method. It's just extremely time-consuming and expensive because of all the hands your invention has to pass through.

The wholesaler gets his cut, the retailer his, until finally you get a small percentage. This isn't bad if you're selling in volume. I'll take a penny a unit for a million units any day. However, large volumes are not necessarily the norm, especially at the beginning of your product's sales cycle. A normal sales cycle consists of four stages: incubation (start-up), maturation (developing sales), plateau (steady sales but no increase), and decline (sales decrease). *All* products follow this cycle.

Obviously, the fewer people the product passes through, the more profit for you. This is why a lot of inventors try to sell their products themselves. I'm all for this, provided you have the time *and sales skills* to do it. If not, get the help you need. Take a little bit less money for a whole lot less headache.

The other methods of selling are:

1. You direct to the consumer.
2. You to the retailer to the consumer.
3. You direct to the consumer *and* through a wholesaler/retailer chain.

Figure 9.1 outlines these four types.

LET'S TALK WHOLESALE

There are wholesalers, and there are wholesalers. You will find distributors, jobbers, mail-order wholesalers, and the regular kind within the world of wholesale.

Wholesale is defined as buying from you at a discount, and then reselling at a markup to other resellers. Let's look at the different wholesale methods.

Full-Facility Wholesalers

A full-facility wholesaler buys from you, sells to resellers, sets up distribution (in this case, movement) of the goods, stores them for you, if necessary,

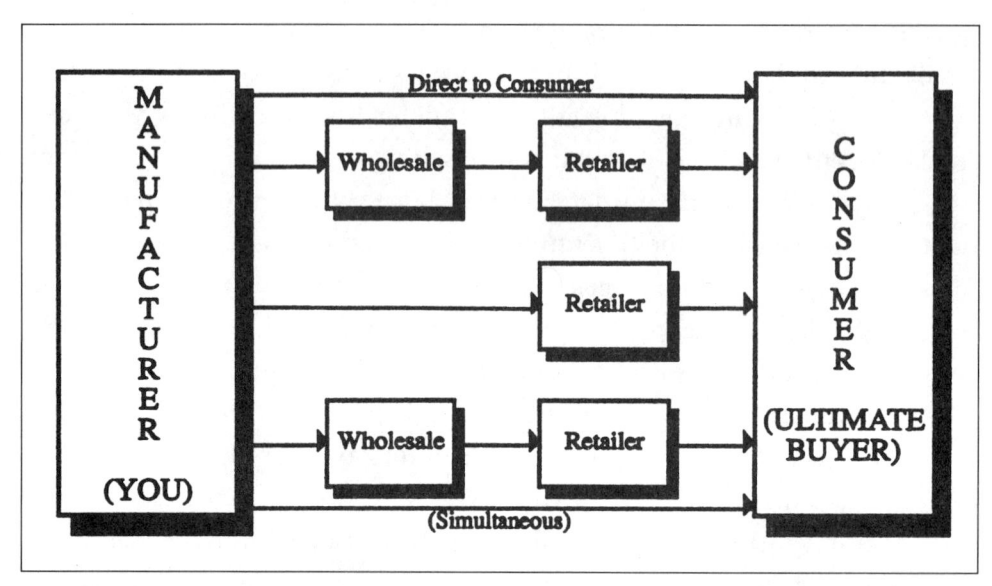

FIGURE 9.1 Distribution Flowchart

helps his dealers and resellers, extends them credit, and does anything he can to facilitate the sale.

Limited-Facility Wholesalers

A limited-facility wholesaler performs very few services other than selling the merchandise. Some give no credit and no storage, and most give very little help to their resellers. (No matter how you sell your merchandise, it is *your responsibility* to train people in the sale of it including the wholesaler's people. As an inventor, it goes with the territory.)

The limited-facility wholesaler is your jobber and mail-order wholesaler. The term *jobber* normally refers to rack jobbers, those "route salesmen" who keep racks of merchandise neatly stacked in groceries and other similar stores. Mail-order wholesalers sell through direct mail and catalogs to the resellers.

Distributors

These are wholesalers who normally do the whole job. If you decide to go this route, make sure they are large enough to produce the volume you want. There are small local distributors and big regional ones. You want to deal with the latter.

Master Distributors

This term is given to a distributor who has exclusive resale rights within a large territory. If someone asks to become a master distributor of your invention, check them out thoroughly. Call a dozen of their clients, and see if they are satisfied. Check with the industry trade association they're involved with, and ask for a status report.

MANUFACTURER'S REPS

This will be the starting point for most inventors. A manufacturer's rep is an independent salesperson or firm that handles several different companys' products, usually in the same industry.

He or she gets a commission, based on volume of sales. Sometimes, in rare occurrences, they get their expenses reimbursed, but since they are independent contractors to you, they should be on their own.

Two directories to help you find rep firms are:

Directory of Manufacturer's Sales Agencies
Manufacturer's Agents National Association (MANA)
23016 Mill Creek Road
Laguna Hills, CA 92653
(714) 859-4040

Electronic Representative's Directory
Harris Publishing Company
2057 Aurora Road
Twinsburg, OH 44087-1999
(216) 425-9000

I'm sure there are specialty rep directories, like the electronics one, for most major industries. Check with your industry association, or with your reference librarian and card catalog.

WHO'S BUYING THESE DAYS?

Thousands of people buy products every minute of the day. They could be buying your invention, *if* you get it to them correctly. As I stated, not every method is right for every type of invention. For the most part, though, the following distribution methods should work for yours.

Selling to the Government

The country's largest purchaser, the U.S. government, through the General Services Administration (GSA) and other agencies, spends billions of dollars every year on products and services. You should try to get *your* fair share. Chapter 10 explains how to do this.

Mail-Order Selling

We discuss mail-order advertising in Chapter 13. While it can be risky if you don't know what you're doing, it's best to start by letting others in the mail order business sell your product for you. By offering your product to mail-order catalog houses, you get the best of both worlds.

Two directories should help you come up with catalog houses to solicit:

Mail Order Business Directory
B. Klein Publishers
P.O. Box 8503
Coral Springs, FL 33065
(305) 752-1708

Lists over 9000 mail-order firms, and what they sell.

Directory of Mail Order Catalogs
Facts on File—Gray House Distributing
Bank of Boston Building
Sharon, CT 06069

Lists thousands of mail-order houses that send catalogs. Also tells whether mailing list is for sale.

RETAILERS GALORE

Thousands of retailers are out there, anxious to sell your product. They just don't know about it yet. (Believe me, they will after you read and apply the publicity principles in Chapter 12.)

How many different types of stores can you think of that might take your invention? The following sources will provide the answers. Your library should carry them.

Department Stores

Directory of Department Stores
CSG Information Services
Lebhar-Friedman Publishers
3922 Coconut Palm Drive
Tampa, FL 33619
(800) 925-2288
$239

Discount Stores

Directory of Discount Stores (8245)
CSG Information Services
Lebhar-Friedman Publishers
3922 Coconut Palm Drive
Tampa, FL 33619
(800) 925-2288
$245

Phelon's Directory of Discount Department Stores, Catalog Showrooms, Drug Chains, & Leased Dept. Operators

Phelon Sheldon & Marsar, Inc.
15 Industrial Avenue
Fairview, NJ 07022
(800) 234-8804
$130

Chain Stores

Directory of General Merchandise,
Variety Chains, and Specialty Stores
CSG Information Services
Lebhar-Friedman Publishers
3922 Coconut Palm Drive
Tampa, FL 33619
(800) 925-2288
$245

Retail Stores, General

Fairchild's Financial Manual of Retailers
Fairchild Books
7 W. 34th Street
New York, NY 10001
(800) 247-6622
$80

LEASED DEPARTMENTS AND RESIDENT BUYERS

There are two other ways to sell your product. Leased departments are departments within a store that it does not own; it leases the space and collects a royalty or commission on the products sold. For example, the Sears eyeglass departments are all leased. If you wanted to sell to them, either contact the lessee directly, or use Sheldon's or Fairchild's to get to the main buyer.

Resident buyers are buyers who work for several different stores, are not employees of the stores, and are located away from the facilities, sometimes out of state—sort of like independent insurance agents. They work for the stores and recommend new products. A good source for you.

SELLING TO THE MILITARY

See Chapter 10.

PREMIUMS AND AD SPECIALTIES

Premiums and incentives are items companies buy to award personnel for a job well done. These can be anything, from travel to automobiles. The largest place for buyers is the annual New York Premium and Incentive Show, held each year.

Ad specialties are those items that get imprinted with company names and slogans. To reach these people, use the *Directory of Mail Order Catalogs* mentioned above, or look in the yellow pages under Advertising Specialties.

TRADE SHOWS

An excellent source of buyers, trade shows are the most misunderstood of all marketing tools. It's easy to sell at trade shows if you know how. Chapter 14 will tell you.

ADDITIONAL RESOURCES

Here are some additional resource books to look up in the reference section of your library. They should prove very helpful:

Direct Marketing Market Place
Hilary House Publishers
980 N. Federal Highway #206
Boca Raton, FL 33432

Lists direct marketers to consumers and industry, service firms and suppliers, and consultants.

In addition, you will find the *Directory of Drug Stores, Grocer's Marketing Guide Book, Sports Marketplace,* and *Microcomputer Marketplace* in your library's reference section.

10

Selling to the U.S. Government

The United States government is the world's largest buyer of goods and services. As a small business operator, you have an ace in the hole with your Uncle Sam. Every federal agency is mandated to buy a certain portion of their goods and services from small businesses. Isn't that wonderful?

I know. You're thinking, "But I have a friend who tried to sell to the government. It was so complicated!" I grant you the process can be complicated, but it can also be rewarding. The problem with most inventors is impatience. If they took the time to investigate the resources available, they would find a wealth of helpful information and people who will guide them *every step of the way.*

In this chapter, we'll briefly discuss how the government buys, who is buying, and what *you* have to do to sell to them. We'll conclude with a section on the different agencies and where to contact them.

HOW THE GOVERNMENT BUYS

Federal purchasing offices buy supplies and services in two ways: formal advertising for bids and by negotiation.

Advertising for Bids

A purchasing office sends "invitations for bids" to businesses listed on its "bidders list" for a certain item. The bidders list is made up of businesses that have told the office they have items for sale, that they can carry out the government's contracts, and that they wish to bid on this item.

In some cases, purchasing offices will want bids from more than those on their lists. In this case, they advertise in trade papers, post notices in post offices, and publicize other ways.

The invitation for bids includes a copy of the specifications for the needed item, instructions for bid preparation, and conditions of purchase, delivery, and payment.

Bids submitted on a proposed purchase are opened in a public bidding process at the office needing the merchandise. Each bid is read aloud, and a copy open for public scrutiny is left for 6 months. At the end of this time, a contract is awarded to the bidder who (1) conforms to all requirements; (2) is considered the most helpful in price, delivery, and other factors; and (3) is deemed competent to carry out the contract.

Buying by Negotiation

Under certain circumstances, government purchases may be made through negotiations with qualified suppliers. The procurement (purchasing) office of an agency asks for price quotes or proposals, known as requests for proposal (RFP). These RFPs are sent to several suppliers to foster competition.

Qualified Products Lists

In certain cases, specific products to be purchased, the invitation for bids, or the RFP will specify that the item is to be on a qualified products list. These items must previously have passed qualification tests. Check with the procurement office of the agency you're dealing with for specifics.

Types of Contracts

In both advertised and negotiated buying, the government generally uses fixed-price contracts.

FINDING BUYERS

It should be obvious that purchases are made from bidders lists, and you must get on them if you want to sell to the government. Before you do this, you'd better know which agencies buy your product types. There are many sources available for this information.

    1. Buy a copy of *U.S. Government Purchasing and Sales Directory*, available from the Government Printing Office (GPO) nearest you. At this writing, the price is $7.00. This directory contains information on how the government buys, who is buying, and other important information.

    2. The Business Service Centers of the General Services Administration (GSA). The GSA is the primary purchasing agent for general use. Its annual disbursement is several billion dollars. Perhaps you've heard of someone getting on the GSA bid list. It's a tremendous accomplishment in itself. Once done, it can lead to constant repeat sales of your invention. Take the time to learn all you can about how the GSA works. Again, the GPO has many books that can help. The GSA acts as a purchasing-agent and clearinghouse for federal and military agencies. It has centers in major cities in the U.S.

    3. The *Commerce Business Daily* is a publication of the U.S. government. Published by the Department of Commerce in cooperation with purchasing agencies, it lists invitations for bids, subcontracting leads, awards of contracts, sales of surplus property, and foreign business opportunities.

    4. The Federal Procurement Data Center (FPDC) is the central clearinghouse for information on most items and services purchased by the federal government.

    5. The Defense Logistics Agency (DLA) is also interested in items that are categorized as general use. Defense Supply Centers, managed by DLA, provide selected commodities for military use.

Preferential Procurement Programs

Federal agencies are required by law to give preference to certain kinds of businesses, such as minority-owned in the awarding of contracts.

Descriptions of the programs follow. (Check the Federal Register at your library for updated information).

Office of Small and Disadvantaged Business Utilization. Public Law 95-507, October 24, 1978, requires each federal agency with purchasing authority to maintain an office of Small and Disadvantaged Business to promote participation of these firms in government procurement. Contact the main GSA office at 18th and F Streets, NW, Washington, DC 20405.

Small Business Set-Asides. This program is authorized by the Small Business Act of 1953. It requires agencies to limit competition on certain contracts to "qualified small businesses." Check with the SBA to see if you qualify.

Socially and Economically Disadvantaged Businesses. This is Public Law 95-507 again. It authorizes the SBA to enter into contracts with other federal agencies. As a result, SBA subcontracts the actual work with socially and economically disadvantaged businesses, defined as any small business at least 51% owned by one or more socially and economically disadvantaged individuals. The law identifies blacks, Hispanics, Native Americans, and Asian-Pacific Americans as those who qualify.

Subcontracts—Small and Small Disadvantaged Businesses. Federal agencies are required to make certain their prime contractors set goals for awarding subcontracts to qualified small and disadvantaged firms. Each prime contractor with a total contract value of $500,000 must do this.

Women-Owned Businesses. Executive Order 12138 requires federal agencies to take affirmative action in support of businesses owned by women.

Vietnam Veterans. There are no legal requirements to award contracts to Vietnam vets, yet. However, federal agencies are actively encouraging them to seek government contracts.

Mandatory Source Programs. If they are offered at competitive prices, the federal government *must* purchase certain goods and services from workshops for the blind and severely handicapped. Further information is available from the GSA.

GSA Purchasing Programs

The GSA purchases goods and services to the tune of billions of dollars each year. It has four major subdivisions: Information Resources Management Service (IRMS), Public Buildings Service (PBS), Federal Supply Service (FSS), and Federal Property Resources Service (FPRS). For further information, contact your nearest GSA office.

IRMS. The IRMS is responsible for automated data-processing and telecommunications equipment, computer software, and services.

PBS. PBS activities include building design, construction contracts, space planning and interior design, office space leasing, maintenance and security of buildings, and several miscellaneous services.

FSS. The FSS is responsible for supplying thousands of common-use items. Their programs include stocking 16,000 common items, vehicles for federal use, travel and freight service, and personal property services.

FPRS. The FPRS is responsible for the disposal of federal real property deemed to be surplus.

GETTING LISTED ON BIDDERS LISTS

After you find out which agencies buy what, ask them to send the necessary forms so you can be placed on their bidders lists (see Figure 10.1). Contact the GSA for their requirements and the DLA for theirs. They will also send you a list of items they are looking to buy.

When you return the form, send it with a cover letter referring to the attached list, and ask which bidders list you will be placed on. Once you get on the list, you will be contacted. You should also regularly contact them.

SOLICITATION MAILING LIST APPLICATION

NOTE: See Instructions on reverse before completion. Type or print all entries. Insert "NA" in items not applicable.

TO: **Western Area Supply Center, Data Automation Division, Attn: Bidder's Mailing List**
P.O. Box 19065, Topeka, Kansas 66619

1. NAME AND ADDRESS	A. PARENT COMPANY *(Name Only)*	EMPLOYER IDENTIFICATION NO.
	B. DIVISION *(Name Only)*	EMPLOYER IDENTIFICATION NO.
	ADDRESS TO WHICH SOLICITATIONS ARE TO BE MAILED	
	C. ATTENTION	
	D. STREET OR BOX NUMBER	
	E. CITY, STATE AND ZIP CODE	

2. APPLICANT'S ADDRESS IF DIFFERENT FROM ITEM 1

3. TYPE OF ORGANIZATION *(Check One)*

☐ A. INDIVIDUAL ☐ B. PARTNERSHIP ☐ C. NON-PROFIT ORGANIZATION

☐ D. CORPORATION; INCORPORATED UNDER THE LAWS OF THE STATE OF ____________

4. HOW LONG IN PRESENT BUSINESS

5. NAMES OF OFFICERS, OWNERS, OR PARTNERS *(Type or Print)*

PRESIDENT	VICE PRESIDENT	SECRETARY
TREASURER	OWNERS OR PARTNERS	

6. AFFILIATES OF APPLICANT *(Names, locations, and nature of affiliation)*

7. PERSONS AUTHORIZED TO SIGN BIDS, PROPOSALS, AND CONTRACTS IN YOUR NAME *(Indicate if agent)*

NAME	OFFICIAL CAPACITY	TEL. NO. *(Include Area Code)*

8. IDENTIFY ITEMS ON WHICH YOU DESIRE TO BID BY MARKING THE CORRECT ITEM CODES AS SHOWN ON THE ATTACHED PS FORM 7429A. PS FORM 7429A MUST BE RETURNED WITH APPLICATION FOR APPLICANT TO BE PLACED ON BIDDER'S LIST.

9. TYPE OF BUSINESS *(See definitions on reverse)*			
	A. MANUFACTURER OR PRODUCER	B. REGULAR DEALER *(Type 1)*	C. REGULAR DEALER *(Type 2)*
	D. SERVICE ESTABLISHMENT	E. CONSTRUCTION CONCERN	F. RESEARCH AND DEVELOPMENT FIRM
	G. SURPLUS DEALER *(Check this block if you are also a dealer in surplus goods)*		

10. SMALL BUSINESS/MINORITY ENTERPRISE IDENTIFICATION CODE *(Check One)*

☐ A — SMALL BUSINESS CONCERN *(But not a minority enterprise)*

☐ B — MINORITY ENTERPRISE *(Whether or not a small business concern)*

☐ D — A BUSINESS CONCERN WHICH DOES NOT QUALIFY AS EITHER A SMALL BUSINESS OR MINORITY ENTERPRISE

CERTIFICATION

I CERTIFY THAT INFORMATION SUPPLIED HEREIN *(including all pages attached)* IS CORRECT AND THAT NEITHER THE APPLICANT NOR ANY PERSON *(or concern)* IN ANY CONNECTION WITH THE APPLICANT AS A PRINCIPAL OR OFFICER, SO FAR AS IS KNOWN, IS NOW DEBARRED OR OTHERWISE DECLARED INELIGIBLE BY ANY AGENCY OF THE FEDERAL GOVERNMENT FROM BIDDING FOR FURNISHING MATERIALS, SUPPLIES, OR SERVICES TO THE POSTAL SERVICE OR ANY AGENCY OF THE FEDERAL GOVERNMENT.

NAME AND TITLE OF PERSON AUTHORIZED TO SIGN *(Type or print)*	SIGNATURE	DATE

Source: U.S. Government Purchasing and Sales Directory, Small Business Administration, 1984.

FIGURE 10.1 Solicitation Mailing List Application

SBA HELP IN SELLING TO THE GOVERNMENT

Here are a few ways the SBA can help you sell to the government:

1. The SBA has Procurement Center reps at major buying centers of the federal government. They will help with contract problems, and they'll offer general advice.

2. Through a program called Certificate of Competency, the SBA provides an appeal procedure if you lose a government contract because the agency felt you couldn't deliver.

3. There are numerous materials on government buying methods, products, and services bought and sold, steps to getting on bidders lists, and other matters. Check your local GPO branch store.

4. Field offices help subcontractors get together with government prime contractors. Ask for a small business contractor specialist.

Small Business Innovation Research

The SBA surveys and monitors SBIR programs of participating agencies. They can guide you through the red tape. Their Technology Assistance Program carries out the requirement to "assist small business concerns in obtaining the benefits of research and development performed under Government contracts or at Government expense."

To get a master schedule of future SBIR solicitations, write:

Office of Innovation Research and Technology
U.S. Small Business Administration
1441 L Street, NW
Washington, DC 20416

R&D OPPORTUNITIES

Here is a partial list of government research and development opportunities, taken from the *U.S. Government Purchasing and Sales Directory*. You can buy the book for $7.00. This list will give you an idea of the vast network of potential buyers the government has. It may also stimulate your creative juices; you may think of a way for your invention to work with a certain agency that didn't occur to you before.

Department of the Army

>Materiel Development and Readiness Command
>Communications Electronics Command
>Electronics R&D Command
>Missile Command
>Human Engineering Labs
>Test and Evaluation Command
>Office of the Chief of Engineers
>Office of the Surgeon General

Department of the Navy

>Office of Naval Research
>Bureau of Naval Personnel
>Medical R&D Command
>Air Systems Command
>Electronics Systems Command
>Facilities Engineering Command
>Sea Systems Command
>Supply Systems Command
>Naval Research Lab
>Ship Engineering
>Air Test Center

Department of the Air Force

>Space Division
>Aerospace Medical Division
>Electronics Systems Division
>Aeronautical Systems Division

Armament Division
Office of Scientific Research

Other Military Agencies

Defense Advanced Research Projects Agency
Defense Communications Agency
Defense Nuclear Agency

Still More R&D Opportunities—Civilian Agencies

Department of Agriculture
Department of Commerce
Department of Education
Department of Energy
Department of Housing and Urban Development
Department of Justice
Department of Labor
Department of State
Department of the Treasury
Environmental Energy Commission
Federal Emergency Management Agency
Department of Health and Human Services
Department of the Interior
National Space and Aeronautics Administration (NASA)
National Science Foundation
U.S. Nuclear Regulatory Commission
Department of Transportation
Federal Aviation Administration (FAA)
U.S. Postal Service

PART V

TO MARKET, TO MARKET

11

Marketing: What It's All About

Part V covers the marketing aspects of your invention—the purpose of this book. As I stated in the Introduction, most invention books deal with the procedures for creating inventions, the prototypes, and the patenting processes. Rarely do they cover the marketing aspects in depth. Because I strongly believe that certain methods are right for inventors, I have devoted several chapters to them.

We will discuss briefly those that *might* be useful to you as an inventor, and we'll cover *in depth* those you should definitely use. The methods with little bearing on your invention (like outdoor billboards) will be mentioned only superficially. Although this approach represents a biased view, it is based on common sense.

This chapter, on marketing, presents the information you need to know about research, competition, and targeting your market.

Publicity can be an entire book unto itself. I devote a long chapter, Chapter 12, on the principles you'll need to expose your invention to the world.

Advertising is one of the most important of the marketing tools. I cover this topic in depth in Chapter 13, with subsections on print advertising, broadcast advertising, mail-order, and direct-mail advertising. These are the types most inventors know about. Yet few of you know how to do them right. You will when you finish the chapter. (If you have questions that just can't wait, contact me at [909] 393-8525, and I'll be happy to answer them.)

Finally, trade shows (Chapter 14) and networking (Chapter 15)—the unsung heroes of the marketing umbrella. The first can be costly, yet extremely effective; the second is inexpensive and highly powerful. You'll learn the secrets of each.

WHAT IS MARKETING?

Marketing is defined as everything you do to promote your business. Everything! This would include advertising, public relations, selling, and all the subclassifications underneath, like newsletters, trade shows, direct mail, sales letters, presentations, and more.

I have identified at least 27 different marketing methods. Some are excellent tools for inventors, others not as good. In order to show where each of these falls, I developed the Marketing Umbrella (see Figure 11.1).

As you can see, there is some overlap. Newsletters *could* fall under advertising, since some newsletters have advertising in them. They could also fall under sales, since they do sell *you* to your clients and prospects. The print and broadcast media also overlap because both carry advertising and editorial.

There is no single correct system. However, we will stick with the umbrella as outlined.

FINDING YOUR TARGET MARKET

The key to smart marketing, and good marketing, in the 1990s and beyond is devoting most of your time to those markets that will afford you the greatest profit. How do you do this?

First, understand that you can't be all things to all people. You must focus on one specific segment of the broad universe we call a market, and work on that until you reach saturation.

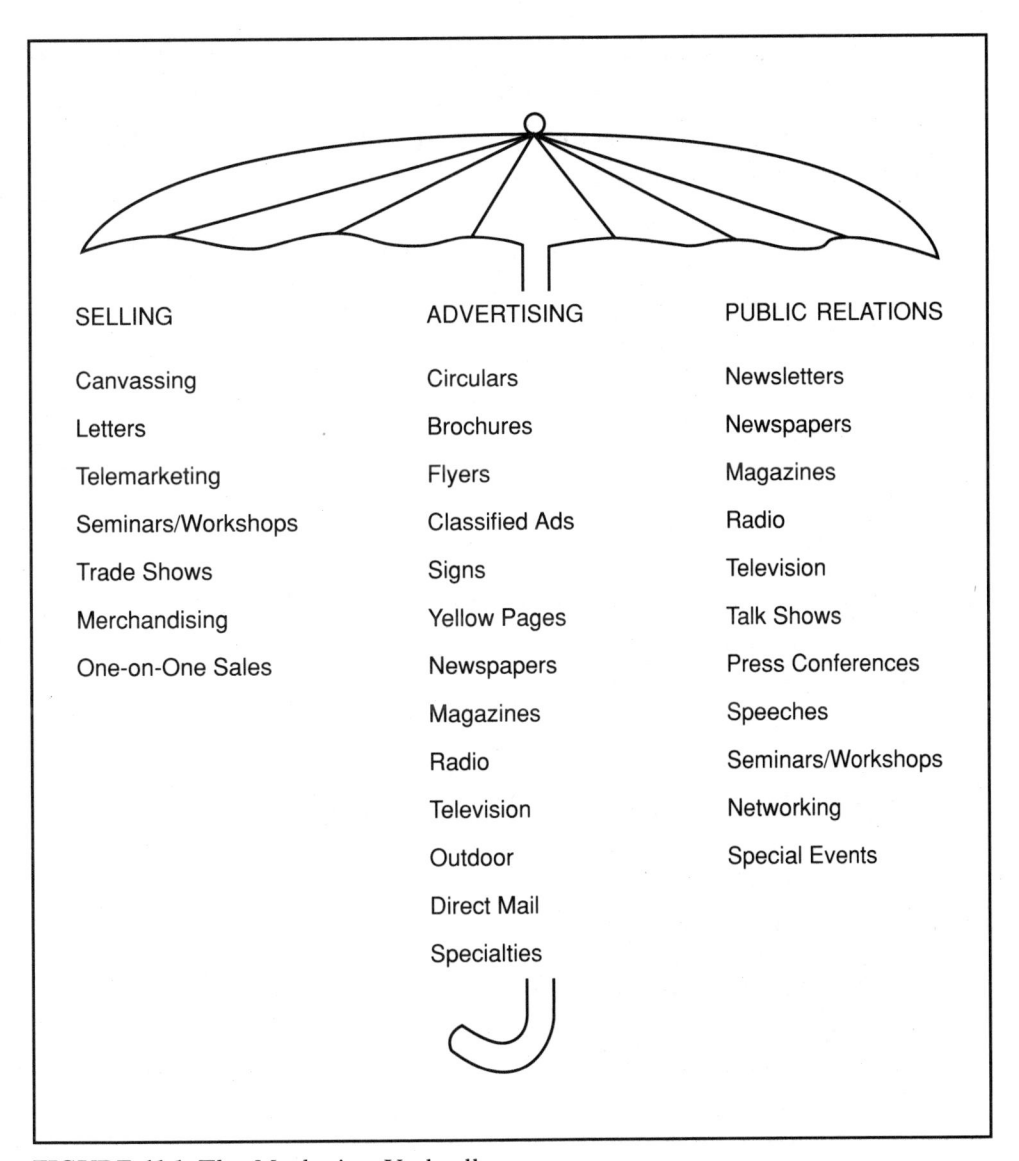

SELLING	ADVERTISING	PUBLIC RELATIONS
Canvassing	Circulars	Newsletters
Letters	Brochures	Newspapers
Telemarketing	Flyers	Magazines
Seminars/Workshops	Classified Ads	Radio
Trade Shows	Signs	Television
Merchandising	Yellow Pages	Talk Shows
One-on-One Sales	Newspapers	Press Conferences
	Magazines	Speeches
	Radio	Seminars/Workshops
	Television	Networking
	Outdoor	Special Events
	Direct Mail	
	Specialties	

FIGURE 11.1 The Marketing Umbrella

For example, let's say you've decided your invention is perfect for children. That's too broad a category. Define it further: What type of children? Where are they located? Who are their parents? What type of income do their parents have? The narrower the target market, the easier it is to penetrate.

Second, once you've found your target market, do everything possible to see that they constantly hear about your invention. Saturate them with all the marketing tools at your disposal.

Third, don't be afraid to abandon ship if it turns out you were wrong. Marketing is scientific to a degree, but is far from infallible. If your target market isn't buying, review your message. If your message is *on target* and they *still* don't buy, perhaps the product is not needed.

STARTING OFF RIGHT

Before we discuss the methods, let's list a few key points. The basic questions to ask yourself are:

- Who can use my product?
- Where can I sell my product?
- How is my product superior to others?
- How does my invention improve lives?
- Is it seasonal?
- What is the competition doing?
- What empty niche is the competition missing?
- How can I capitalize on the empty niche?

Who Can Use My Product?

This is a most important and basic question. Answer this one first. Ask yourself who can and would use it, and make a list. Then rank the list from high to low. Research the top three target markets in depth.

Where Can I Sell My Product?

Determine geographically where your product is sellable. Overseas selling requires different marketing strategies than domestic selling. Selling in the Northeast U.S. may be different from selling in the Southwest.

How Is My Product Superior to Others?

This is a key question. In marketing, we call it the *unique selling proposition (USP)*. It answers the question, What is the one thing that makes my invention or product unique from all others like it? Is it quality of parts, is it function, is it cost, is it composition? Whatever makes yours unique, and better, *that's the key*. Focus on that uniqueness, and make it your number-one selling statement. Let your target markets know constantly what it is and why they should have it.

How Does My Invention Improve Lives?

This one goes along with the previous question. Once you know your USP, you need to show how it improves people's lives. Do that, and you've got a winner.

For example, if you'd just invented the electric can opener, you might say your USP was functionality—ease of use. And how does this improve lives? Of course! The ease of use saves time, saves scraped knuckles, and—most important—saves money! See how this works?

Is It Seasonal?

If your invention is marketable only in certain seasons, its shelf-life will necessarily be short. You will therefore have to market it the fastest ways possible, like print advertising and mail order, which are more costly than slower methods like direct mail and publicity.

You will also have to raise the price, because not everyone will be enamored of your particular seasonal product. Some like to ski, and some like to sun. Even though you have a narrow market, your costs could be greater than if you had a broader market.

What Is the Competition Doing?

If you don't know who your competition is, you're in trouble. A smart marketer knows who the players are, and how they play the game.

What Empty Niche Is the Competition Missing?

If there is one, how can you find it? More importantly, how can you capitalize on it? To see how this works, look at Figure 11.2.

Type of Potential Client	Under $25 per class	Under $50 per class	Over $50 per class
Beginners	RCC CSUSB Adult Ed.	Software Station Dean's Computes ABC Computers	Computerland Microage Businessland
Intermediate	RCC CSUSB Adult Ed.	Software Station Dean's Computes ABC Computers	Computerland Microage Businessland
Advanced			Computerland Microage Businessland
Educators	RCC CSUSB Adult Ed.	Software Station Dean's Computes ABC Computers	
Small Business	RCC CSUSB Adult Ed.	Software Station Dean's Computes ABC Computers	Computerland Microage Businessland
Corporate			
Home Users	RCC CSUSB Adult Ed.	Software Station Dean's Computes ABC Computers	Computerland Microage Businessland

FIGURE 11.2 Competitor Analysis Grid

This competitor analysis was done for a previous client of mine. We asked her to name the top ten competitors in her area, computer instruction. We then designed the matrix grid to show type of potential client, or target market, and pricing structure. We then asked her to rate them based on targets and money charged. The results are shown in the figure.

What we found were six potential holes or empty niches in the marketplace she could have captured with no initial fight. Any one of them would have put her in the envious position of having "been there first." You can use this matrix for any type of similar comparison.

How Can I Capitalize on the Empty Niche?

This is what the pros call *positioning*. Remember the saying, "Find a need, and fill it"? The new marketing—positioning—says, "Find a need that no one else has capitalized on, get there first, and hang on like a tenacious bulldog."

David Ogilvie, guru of advertising and marketing, put it best when he said, "In my 30 years of advertising, one thing that made the winners was their positioning."

Here are some examples. Avis and its slogan, "We try harder." Or 7-Up, the Uncola. Again, what do you have that's unique, that sets you apart? Use your unique selling proposition, (USP) and hit hard at the holes you found in the marketplace. You'll come up smelling like a rose.

THE IMPORTANT THREE QUESTIONS

Now that you're thinking like a marketer, it's time to expand your knowledge further. Marketing requires developing a plan of attack. But before you attack, you must know yourself, the enemy (your competition), and how the enemy thinks and acts. You must know your own strengths and weaknesses. And you must also know who you want to woo (your customers).

Think of marketing as you would a romantic movie. The hero is you, the inventor. The heroine is your target market. And the other suitors are the competition. You and your competitors will put your best foot forward (your USP) to win the heroine's heart. But first things first.

Question 1: Who Are You?

Know what you do best, and what you like to do. Know your own strengths and weaknesses, and capitalize on the strengths. Know your own capabilities, and get help where needed.

Question 2: Who Is Your Customer?

This is the target market aspect. Who do you want to sell to, and where are they? How large a market are they? Are they temporary, or permanent?

What do they want from you? See the Customer/Product Analysis in Figure 11.3.

Question 3: What Is the Marketplace Like?

What's out there? Is the competition breathing down your neck? What are their strengths that could hinder you? What are their weaknesses you could overpower? Who's playing the game? Is there room for all the players? See the Competitor Evaluation Checklist, Figure 11.4.

Ask the following questions for each major product group and for each major customer segment. But don't just ask the questions; find the answers, investigate past trends and likely future trends, and carefully consider the implications.

- Who is the likely customer?
  - Where does s/he live?
  - What is his/her age?
  - What is his/her income?
  - What is his/her level of education?
  - If the likely customers are other businesses . . .
  - What is there size?
  - What are their normal, expected discounts?
  - Who are their customers?
  - Who normally makes the purchasing decisions?
  - What are their purchasing procedures?
- How many potential customers are there?
- What are customers' quality expectations?
- Are customers likely to perceive a purchase risk?
- What needs does the product satisfy? (Think carefully about this question.)
- How is the product used?
- Are there alternative uses?
- Is the product a necessary or discretionary item?
- How many units is the customer likely to buy?
- How often will the customer buy?
- When will the customer buy (day of the week, time of day, season of year, etc.)?
- If the product is a service, does the customer need to be present when the service is provided?
- Where would the customer learn about the product (e.g., friends, business associates, TV advertising, yellow pages, etc.)?
- Who actually buys the product (e.g., Mom, purchasing agent, other)?
- Who influences the decision to buy (e.g., kids, engineers, etc.)?
- Why should the customer buy your product and not the competitor's?
- How much prior knowledge does the customer have about the product (i.e., is it truly a new innovation)?
- How much would the customer likely pay for the product?
- How sensitive would she or he likely be to price changes?

Source: Charles L. Martin, *Starting Your New Business,* Crisp Publications, Inc. 1988.

FIGURE 11.3 Customer/Product Analysis

> Use the following checklist to evaluate the competitors your business is likely to face. Be careful not to define your competitor too narrowly. For example, if you're considering opening a new movie theater, your competition isn't limited to other theaters. You'll also be competing with VCR rentals, bowling centers, and other forms of entertainment for the consumer's discretionary dollar.
>
> - Who are your potential competitors?
> - What are their strengths? Weaknesses?
> - Who are the customers of each competitor?
> - Why might a consumer buy from them instead of you?
> - What is the approximate sales volume of each major competitor? Are there any significant trends in sales?
> - What is the market share of each competitor?
> - What is the cost structure of each competitor? Lower overhead? Higher?
> - What is the pricing structure of each competitor? Below your products and/or services? Higher?
> - Do competitors enjoy support from a strong franchise parent company?
> - How do competitors promote their products, services?
> - What are the distribution arrangements for major competitors?
> - Who are the suppliers of each competitor?
> - How is each competitor positioned? What is the mental image that comes to the consumer's mind when she or he thinks of each competitor?
> - Are there known potential future competitors not currently operating in the industry? If so, who?
> - What are the management strengths and weaknesses of each major competitor?
> - Are the competitors well financed?
> - How committed is each competitor? Will competitors be forced to vigorously compete after you enter the market?
> - Are future technological developments likely to alter your competitor's product line and/or mode of operation? Are they better prepared to adapt to change, or are you?
> - How does each competitor's product line rate in terms of breadth and depth?
> - How do competing products rate in terms of quality, size, appearance, durability, packaging, etc?
> - What are the credit terms of major competitors with customers/suppliers?
> - Do competitors stand behind their products? How do their warranties rate?
> - What sort of auxiliary services do competitors offer (e.g., gift wrapping, installation, delivery, maintenance and repair, etc.)? Are customers charged separately for these services?
> - Do competitors own any patents or any exclusive distribution rights that would affect your market entry?
> - If competitors are retail stores or service businesses, how do their physical facilities rate in terms of layout, decor, cleanliness, parking, convenience of location, ambiance, etc.?
> - What are their hours of operation?
> - Within the community, how saturated is the market? In other words, is there room for your new business?

Source: Charles L. Martin, *Starting Your New Business,* Crisp Publications, Inc., 1988.

FIGURE 11.4 Competitor Evaluation Checklist

THE MISSION STATEMENT

Your business's mission statement is one of the first things you must analyze for the purpose of marketing. From the mission statement flow all the other questions, answers, and ideas that will help you determine what you will market, and how and to whom. It is your reason for being in business. If you have not yet written a mission statement, do it now.

Here's an example:

> The mission of The Franklin Group is to create awareness and understanding of small business problems and to help find solutions to those problems. We do this through various media methods, such as books, tapes, videos, and seminars. For a complete catalog of products and services, write to P.O. Box 2667, Chino, CA 91708-2667.

DOING YOUR OWN MARKETING

Although I believe you should be able to do the bulk of your own marketing, you may need help with certain aspects. It never hurts to get outside opinions. You can't do everything yourself. But try to do as much as possible on your own. The more you know, the more you grow.

If you need some help, such as a direct-mail specialist, advertising copywriter, or a freelance graphic designer, the following information should prove useful.

FARMING IT OUT

When farming out part of your marketing, you need to know what to look for in hiring part-time professional help. The three ways to subcontract are through consultants and independent contractors, advertising and public relations agencies, and sales reps.

Consultants and Independent Contractors

These days, anyone can hang out a shingle and say they're a consultant. The more people out there needing help, the more lucrative consulting seems, and the more people become consultants. However, there are consultants, and there are consultants.

A true consultant is totally independent and will not push you to use any single marketing method, or vehicle, until he or she is certain it is right for you. The consultant holds no allegiance to any publication or media. A true consultant does the following:

1. Weighs all the factors carefully before taking on your assignment.
2. Knows where to look for answers and is well versed in market research.
3. Tells you up front if he or she cannot help you. (Compare this with the so-called inventor's marketing companies that say every invention is marketable and are really conning you for your money.)
4. Belongs to one or more associations in his or her field of specialization.
5. Specializes in certain marketing aspects. Like yourself, he or she cannot be all things to all people.
6. Charges by the hour or on a retainer fee, not by a percentage of your gross sales.

Advertising and Public Relations Agencies

These days, many advertising agencies also do public relations. Chapter 12 shows you how to do your own PR. Unless your project is so large that it is overwhelming, it will not be necessary for you to retain a publicity agent. (Good agencies do, however, have good media contact files you may want to exploit.)

Advertising is more sophisticated than PR. If you lack skills in copywriting and ad design, it might be a good idea to explore hiring a part-time ad agency or freelancer.

What should you look for in an agency? The mark of a professional agency is that they care more for you and your invention than they do about winning awards. Lots of agencies boast that they've won Clios (the Oscar of the agency business) and other awards. But *awards don't sell inventions—good marketing does.*

Finding a Good Agency

- Ask for recommendations from other inventors and friends.
- Ask at inventors' clubs. (See Chapter 15.)

- Check the local business-to-business yellow pages. (Contact at least six, and set up interviews.)
- Call your local Advertising Club Referral Service. (Call an agency from the yellow pages, ask if there is a local club, and get the referral number.)
- Read your regional edition of *ADWEEK,* a trade magazine. Many freelancers advertise in the classified section.

Sales Representatives

Here we're talking about manufacturer's reps. Sales is a function of marketing, and one of the many methods open to you is to hire a manufacturer's rep.

- There are several associations of reps in the United States. One of the largest is MANA, Manufacturer's Agents National Association in Laguna Hills, CA (714) 859-4040.
- To find other rep associations, specifically by industry, research the *National Trade and Professional Associations Directory* in the library.
- Ask other inventors, or inventors' clubs, for referrals.
- Check out classified ads in trade magazines.
- Call other manufacturers in your invention area from the Thomas Register. Ask them for the name of the most reliable rep firm they know.

THE INTERVIEW

Here is a list of questions you should ask before hiring an agency, a freelancer, a sales rep, or a consultant:

1. Are they inventive?
2. Are they interested in my invention and company?
3. Do they understand my industry?
4. Who are their other clients?
5. How long have they been in business?

6. Is my business large enough to keep them interested?
7. Do they work for my competitors?
8. Who will do the work?
9. How do they charge?
10. How should I pay?
11. What does each aspect cost?
12. Can I get a written estimate?

Are They Inventive?

Don't hire people who do the same thing for all clients. How creative are they? How different is their work for other clients? Ask to see samples. Get at least six referrals, and call them. Ask if these clients are (or were) satisfied. If yes, why? If not, why not? You *must* check them out.

Are They Interested in My Invention and Company?

People who take on your project just because they need the work are not going to be very responsible to you. They must show a genuine interest in your invention, *and* in the growth of your company.

Do They Understand My Industry?

Many agencies specialize in certain industries. Some will only deal with real estate companies, others with the electronics field, others still with toys and games. Try to pick someone who has knowledge and expertise in your particular field of invention.

Who Are Their Other Clients?

This goes along with the previous question. Who are their other clients, past and present? Why did they lose prior clients? How do they relate to their current ones? Are there direct competitors they now work for? Will they drop them to work for you? Don't hire someone who currently deals with a direct competitor.

How Long Have They Been in Business?

You don't want a rookie fresh from a trade school, or a newly opened agency, unless they have credentials to prove their competence. On the flip side, a seasoned professional agency probably won't think your account is big enough for them. Try to find someone with at least 3 years of experience.

Is My Business Large Enough to Keep Them Interested?

To quite a few agencies, you will probably be considered small potatoes. Some inventors have called agencies, only to find out that unless they billed *at least* $15,000 per month, the agency wasn't interested. A solution might be to hire a designer, copywriter, or account executive from the same agency on a freelance basis. Lots of them look for ways to supplement their salaries. Check the *ADWEEK* classified section.

Do They Work for My Competitors?

As stated before, don't hire anyone who currently works for a direct competitor. While reputable pros will not divulge client information, you cannot afford to take the chance. If they agree to drop competitors for your business, they might do the same to you later on.

Who Will Do the Work?

Make sure the person you're interviewing, and whom you feel comfortable with, is the one doing the work. Even in small firms, very often the owners will meet with you to sell you on their services, then turn the work over to someone else. Know this up front. Ask to meet the staff assigned to your project before you contract with the firm.

How Do They Charge?

Will you be on payment terms? Net 10 days? Net 30 days? Cash up front? Hourly or retainer fee? Just what does the retainer fee mean? What constitutes an hour? What do you get for your money? Don't be afraid to ask questions—it's your money and your company!

How Should I Pay?

I recommend that for short-term projects, you pay in thirds on the system. This means you pay one-third as a down payment to show good faith, one-third halfway through the project, and one-third at the end of the project. This system protects you and gives you leverage if you're not happy with the work. For longer projects, use monthly retainers, where you pay month by month. Know in advance what you'll get for the retainer.

What Does Each Aspect Cost?

You *must* be aware of what each segment of your marketing costs. Each element must be broken down for your inspection. For example, if market research will take 100 hours, this should be broken down into library research, designing survey forms, collecting the data, analysis of data, etc. The more defined you get, the better off you are.

Can I Get a Written Estimate?

All building contractors give written estimates of their work before they begin. There's no reason why you shouldn't get one for the work someone will do for you, either. Make sure each aspect as discussed above is itemized and priced.

ON TO THE TACTICS

Now that you've got your marketing mind in gear, let's go on to a low-cost, high-result tactic—*free* publicity!

12

How to Get Free Publicity

One of the best ways to promote your invention is through free publicity. Publicity can move mountains, get a bill passed or killed in Congress, elect a U.S. President, or sell your invention.

DEFINING PUBLICITY

Publicity is using various communication tools to persuade the media to print your story or broadcast news about your invention or company. It doesn't cost you airtime or ad space. In this respect, it is free.

But publicity is never totally free. There is no such thing as a free lunch. You will have to have some type of budget for stamps, envelopes, phone calls, reproduction of releases and photographs. And you will have to *pay your dues* in sweat equity. What you don't spend in cash, you may spend in time.

Why Is Publicity Important?

Getting good publicity for your invention is the easiest way to gain credibility. You may even get some orders.

Publicity is important to the media, too. Local activities are news. Let's face it, being a reporter or editor is a very hard job. They can't be everywhere, and you are their eyes and ears—their source of information. They need you as much as you need them.

MEDIA MYTHS

As a former newspaper editor and publisher, let me allay your fears: The awful things you hear about the media are simply not true!

Myth Number 1: Media Are Out to Get You

Nothing could be further from the truth. I have known media people from large concerns to small operations. Most are warm, sensitive human beings with too little time and too much to do. Media people don't have time to plan arduous campaigns to hurt you, and that's not their intention anyway.

They will, however, remember if you wrong them. So be careful! They won't take time to get even, but you'll never get their respect or help again. Know the rules of the game, and you'll do fine.

Myth Number 2: Media Play Favorites

It is true that many reporters and editors have their favorite people from whom they accept information. The way to understand this is to ask why.

Most media people have a ranking system, from the lowest of the low (those who take and never give) to the highest—what they call their "resident experts."

The reason the press deals repeatedly with the same people is that they feel comfortable with them. These people have become excellent sources of honest, factual material. Editors and reporters can trust them.

If you want to be at the top of the pecking order, become a trustworthy source of information. Give it freely, whether you want something back or not.

Myth Number 3: You Have to Bribe the Media to Get a Story

Unfortunately, in some publications or broadcast media, this may still be true. For the most part, however, because of the infamous payola scandals of the early 1960s, bribery is almost extinct.

WHAT MAKES A STORY SELLABLE?

The press will take your story if it meets certain criteria:

1. It must be timely.
2. It must be newsworthy.
3. It must have human interest.
4. It must follow correct procedures for presentation, style, and delivery.

Timeliness

If you send in a news release two days before the event, don't expect it to get printed. Air time and print space are at a premium like never before. Your material must reach the editor in a timely fashion for use, and it must arrive at least a week before your event.

Newsworthiness

There's hard news, and there's soft news. Hard news is that which the media *must* report, such as a fire, tornado, or a political event. Soft news is what you will be sending—your product releases, company expansion plans, etc. The closer you can make your soft news be to hard news, the easier it will be to get in the media.

Human Interest

Human interest is the key. Readers, viewers, and listeners are human beings, and editors and news directors look for stories with the humanity aspect. Put humanity into your PR, and you'll come out a winner.

Correct Procedures

There are certain forms, procedures, and protocols that the media live by. Knowing these, you are certain to leverage your chances for getting free publicity. In the next section, on establishing relationships with the media, we'll explore some of the correct ways to reach them.

ESTABLISHING RELATIONSHIPS WITH THE MEDIA

If you can, personally take a trip down to your local paper and introduce yourself to the business or science editor. (Who you deal with will be determined by your product and industry.) Use this approach:

1. Determine the best time to visit.
2. Bring basic material with you (keep it under cover).
3. Don't sell the story on the first visit.
4. Follow up with a phone call about some other local story that doesn't relate to your business.

If an editor is unreceptive, simply say you came down to meet him or her and would like to leave your business card. If the staff needs any information on the industry you deal with, you'll be glad to help. Say thank you, and leave.

When you get back to your office, send a note thanking the editor for taking time to meet you. Little courtesies start you off right.

If you cannot see the editor and are rebuffed by a gatekeeper, be polite. Say thank you anyway, take out your business card and press release, and ask that they be put on the editor's desk. Say thanks again, and leave. (You want this gatekeeper on your side.)

When you get back to the office, call the editor later in the day, apologize for the inconvenience, and say you had stopped by to deliver some information. Ask to schedule an appointment if the editor finds it newsworthy. Send them a thank you note, no matter what they do.

The Long-Distance Relationship

Most inventors want to hit the "big papers" so fast, they forget the impact of the local press. The previous discussion was obviously for dealing with local papers, especially weeklies.

But what happens if you want to sell your story to national trade and consumer magazines, and national newspapers? You can't possibly hand-deliver each release around the country.

This is where you must mail each release out. However, as long as you don't send a shotgun, or blanket release and instead tailor your release to each publication, it's almost like walking into the publication's office and chatting with the editor.

Once the proper release is sent, it's time to follow up with a phone call. (Only do this for long-distance magazines and newspapers.) Be polite and brief. Get to the point. Don't ask if they received it, or if they intend to use it. Try, instead, "I'm new to this. Did I send you the information correctly? How would you like it done in the future, so that it has a good chance of getting in?"

Become Their Local News Source

Once you've established contact with the correct media person, whether the relationship is local or long-distance, you keep the connection alive. Send information on stories their readers might find interesting. Send them information that might help them. Pretty soon, you'll find yourself becoming a source for them. That's what you want.

Resident Expert—Your Ultimate Goal

Your ultimate challenge, and goal, is to become an editor's resident expert in your industry. In other words, whenever they need someone to answer a question about your product's industry, or about inventions in general, your name surfaces in their minds. You can become a resident expert to an editor through the various methods you will learn, such as relationship development, article writing, and possible column assignments.

The Media Contact File

You should develop a Media Contact File. On the back of each card in the file, keep good notes of your conversations with every editor. Over several months, you'll see an increase in your contacts and in your relationship. Pretty soon you'll "red-flag" that card. That's the day you become that editor's resident expert. Then get out the champagne!

Your Media Contact File is like the ones professional public relations agencies use. Only this one is customized by you specifically for your invention's publicity campaign.

Get a 5 × 8 file box and a couple of packages of 5 × 8 cards. Take blank divider cards, and label them as follows:

- Main headings: Print Media and Broadcast Media
- Print subdivisions: Magazines—Consumer and Trade; Newspapers—Dailies and Weeklies; Newsletters.
- Broadcast subdivisions: Radio—Commercial and Public; Television—Commercial, Public, and Cable; Talk Shows.

Include the following information on each contact card:

Name of publication/station/show/service
Name of contact person
Street address
Mailing address
Telephone number
Fax number
Deadlines
Photo requirements
Circulation/Broadcast range
Time of pertinent columns/broadcasts
Publicity materials accepted

Include on the back of the card:

What was sent, when, and to whom
What was aired or printed, and when

Contacting the Right Person

Without the proper person to contact, all your research and preparation will be wasted. You need to know who the people are who can influence whether your story gets told.

Newspapers. At a newspaper, contact the section editor or beat reporter. These are the people who specialize in certain areas of expertise: sports, small business, high technology, etc. Try first for the editor; he or she has the clout to say yes or no to you. Then try the reporter. (On small weeklies, your starting point, this may be one and the same person.) Read the columns these people write; if you want to have your material accepted, the fastest way is to write like they do. Know their style, and what type of ideas excite them.

Magazines. At magazines, check the masthead for the senior editor or features or articles editors, if you intend to send articles on your invention. In this case, you are almost acting like a freelance writer. And who knows, you may be lucky and get paid for promoting your own invention! Wouldn't that be sweet?

In sending releases to magazines, make sure you send them to the appropriate department editors (e.g., science, women's, lifestyle, etc.). Do not try to contact people listed as contributing editors; they usually are freelance writers and don't have the authority to okay your story.

Radio and TV. For radio and television, your contact person will be the news director and, prior to an event just happening, the assignment editor. The news director will pass your release on to the appropriate editor, if there is one for your subject. If not, the news director will keep it.

If they decide to use your story, it will likely be sent on to an assignment editor for assigning to a reporter. Your contact with the assignment editor will be just prior to any event you are involved in. Call up and ask politely if they intend to send a reporter to cover the unveiling of your new product. If they say they didn't receive the release, fire off another one immediately! I've saved a few press conferences from disaster by using this method.

TACKLING A PUBLICITY PROJECT

The Publicity Project Tracking Sheet (in the Appendix) is a checkoff list, designed to aid you in tracking your progress through a publicity event, or story idea. You start the process by making a list of magazines that might take your releases and articles. To get ideas, ask yourself some questions about your potential customers. If your invention is a toy, for example, "Where would people read about children's toys?" You should come up with a list of at least six consumer magazines. And you'll probably realize that there is at least one trade magazine for your industry subject. Virtually every invention subject has a trade magazine. Consult the various directories to find out what they are. (See the descriptions below, beginning with General Media Directories.)

Approaching the Market, Researching the Media

Ask yourself whether you want to approach your target market geographically or demographically. If geographically, pick the area you intend to saturate with your publicity campaign, and analyze what the target audience would read in that area. If demographically, determine your audience profile from your basic market research, and assess which media they will use.

Next, acquaint yourself with the various media directories, so that you can develop your Media Contact File. Listed below are the main directories with which you should be familiar. There are over 200 media directories for the U.S. alone. For very specific information, such as Hispanic or black media, use *Gale Research's Directories in Print*.

General Media Directories

Bacon's Publicity Checker. This is the premiere sourcebook for publicists. It lists daily and weekly newspapers all over the country, as well as radio and television stations. The newspaper volume, for example, is divided geographically, then further subdivided into dailies, weeklies, and multiple publishers.

Let me clarify that last statement. Some publishers put out several area weekly papers, or they have several magazines. They are listed in

Bacon's as multiple publishers. On newspapers, one editor generally does the work for all the publications, because the bulk of the paper contains the same news. Only the community sections change. So send your release or article to only one editor. (For magazines, this is different. Multiple magazine publishers have different staffs for every publication they print.)

Working Press of the Nation (WPN). This is an excellent five-volume set, with a separate volume each for newspapers, magazines, radio, television, and internal publications (company newsletters). Add whatever information you did not get for your cards from Bacon's, using *WPN*.

One of the nice things about *WPN* is its Publicity Material Table at the bottom of each page. Within each listing, there is a notation line labeled "Publicity Mtls:," indicating which publicity materials or tools they take. By matching the grid at the bottom with the numbers in each individual listing, you can tell what they take. (All take releases but may not say so.)

Standard Rate and Data Service (SRDS). This thirteen-volume set of media directories is very valuable. One volume is on consumer magazines; another on trade magazines. They are indexed by subject matter, so your search should be easy.

Gale's Directory of Publications. This lists the various publications by city with their circulation figures, certainly something you should put on your contact card. You will use this information later when you get some press coverage and want to send clip sheets (see page 156).

Gebbie's All-in-One Press Directory. Personally, I don't like Gebbie's. It has no information I can't get from all the other sources in my library. But since many people use it, I feel obliged to include it here.

Writer's Market. This is a sourcebook of over 4000 magazine markets. Normally used by freelancers writing for pay, it will give you valuable information for your Media Contact File.

Newsletter Directories. Newsletters can be an excellent source of media exposure for you. There are industry newsletters, internal house organs, corporation newsletters, etc.

Two excellent directories are the *Oxbridge Directory of Newsletters,* and the *Newsletter Directory* by Gale Research. And don't forget *Working Press of the Nation, Vol. 5,* for additional names. Check the reference room at your library.

Radio and Television Directories

The best resource is *Broadcasting/Cablecasting Yearbook.* It's an excellent guide to all the television, radio, and cable stations in the United States. If you intend to eventually self-syndicate your own radio and television cassette releases (this is expensive and very sophisticated), this book will help.

Also, *WPN* has a volume each on radio and television stations, as does *SRDS.*

Association Directories

Another good source for your Media Contact File is association directories. Gale puts out the *Encyclopedia of Associations* in a three-volume set. Volumes 1 and 2 are national listings, and volume 3 lists regional associations.

The problem with all the directories is that they're dated. Most come out once a year. And believe me, editors and reporters move around like you and I turn on the water faucet. And in some directories, like *SRDS,* the publishers send in the information, so what's printed may be inaccurate. Worse yet, many publications do not send in their information to all the major directories. Therefore, you have to consult several directories to complete your Media Contact File.

Look through the directory index for the subject area of your invention. Then write to these associations, asking if they have a magazine or newsletter they distribute to their members. If they do, you have another possible source for publicity.

The Selection Process

Now that you know where to find the media resources, pick the magazines, newspapers, and radio and/or TV stations that seem likely to accept your material, based on matching their audience with your target audience.

A chart I developed to sort the data is the media analysis grid (see the Appendix). Write the name of each media vehicle at the top. (Use a different sheet for each media type—radio, TV, magazines, newspapers.) List the publicity materials taken along the side, as in *WPN*. Check off from your contact cards the appropriate information, and add up the score. If twelve of your vehicles take news releases and only five take articles, do the releases first, and leverage your chances. Then move on to the next most-frequent tool, and so on.

Obtain Samples

You *cannot* write good news releases or articles unless you know about the media vehicles for which the written materials are intended. The best source of this information is, of course, the media themselves. Call all those to which you intend to send information, and ask for a media kit. Make sure they include reader, listener or viewer profiles. Ignore the puff pieces—the "we're the #1 radio station" material.

Study what they send you thoroughly. Read the articles by the editors or reporters you intend to contact. Watch and listen to the news on the stations you will reach. The more you know about them, the more you can write in their style, and the easier it will be to get them to accept your material.

Next, select the top six in each media category (radio, television, magazines, newsletters, and newspapers), and target them for a full-fledged campaign!

Preparing the News Release

Before writing the release, you may need additional information. The card catalog and *Reader's Guide to Periodical Literature* are next.

The Card Catalog. Use your library's card catalog for fact finding. Always include facts in your releases and articles. In fact, articles are made up of facts, quotes, and anecdotes. How they are proportioned determines a writer's style, and subsequently, that of a publication.

Use the card catalog to look up subject matter on your industry, and integrate good, solid facts into your media releases and articles.

Reader's Guide to Periodical Literature. The *Reader's Guide to Periodical Literature (RGPL),* also in your library, is a sourcebook for previously written magazine articles. This book can be the basis of your research, and can save you countless hours working on articles that have already been written.

Let's say you intend to send a news release about your invention. (As an example, let's use a new toy and call it the schlumper.) Before you write the release, it would be a good idea to know several things:

1. Has anyone written about the schlumper before?
2. Is there another toy like it already?
3. What magazines would carry articles on it?
4. What are the trade magazines for toys, and how do they cover new products?

Make a list of all possible magazines that would take your release or article. Then check under the appropriate headings—in this case, toys and schlumpers—for the last 3 years. If you find an article, chances are the magazine will not repeat itself. (Magazines generally recycle article ideas once every 3 years.) Drop that magazine from your list temporarily, or give your release or article a different slant.

For those magazines that have not published articles about your subject, find related articles in the *RGPL* and study them. You'll have a far better chance of getting yours printed.

Sending the Release

Send a cover letter or note along with *anything* you send to the media; it demonstrates professionalism. Don't use a postage meter or a bulk-rate permit; this reveals a shotgun mailing approach, one that editors hate! Don't use computer-generated address labels; prepare each envelope individually.

What should you send? For print media, send the release, cover letter, product photograph, and a fact sheet about the product. For radio, the same. (The photo may intrigue the news director enough to grant you an interview.) For television, send a slide instead of a photo, along with the rest.

Tracking the Results

If you don't know what's being printed or broadcast about your invention, what's the sense of doing it? You must keep track. The Appendix includes a simple media tracking sheet for following your results.

Budgeting and Scheduling

Even public relations needs a budget. It takes money to reproduce articles and releases. You may need a press kit, and photographs cost money. You won't need a very large budget. Plan to spend roughly $500–$1000 for an initial news release for a major invention; $250–$500 for a smaller product.

How should you budget? The old method was to spend money evenly, every month, to get a release out. This does not take into consideration special events, press conferences, or anything else you may want to do.

I recommend you budget at least 5% of your total promotion budget for publicity. Put most of your money into one or two major promotions per year, and parcel out the rest equally for "keeping-your-name-in-front-of-the-press" news releases.

Basic Strategies for Invention Publicity

There are several publicity strategies that fit well into invention marketing. These include news releases, interviews, and talk shows; product update campaigns; technical articles for trade magazines; and product roundups, where your invention is one of many featured within a consumer or trade journal.

Other Vehicles to Use for Free Publicity

What are the different vehicles that will work for you with these strategies? Start with print media, then move on to broadcast media.

Newsletters. Newsletters are inexpensive to do, provided your story appears in the right company publications.

Speeches and Lectures. These are relatively inexpensive and easy to do. Use the *Encyclopedia of Associations' Regional Editions* in the library to find lists of local chambers of commerce, Rotaries, Kiwanis Clubs, Elks Clubs, and others to speak to.

Seminars and Workshops. Although these require planning and intense preparation, they can yield great contacts and possible buyers. (For a complete guide to putting on seminars, read Gordon Burgett's *How to Set up and Market Your Own Seminar* Communications Unlimited, P.O. Box 6405, Santa Maria, CA 93456).

Networking. Networking is an obvious vehicle for free publicity, and a skill we already talked about. It's so important that I've devoted an entire chapter to networking organizations, Chapter 15.

Talk Shows and Press Conferences. These are the most sophisticated and impressive tools you can use. The person to contact for talk shows is the program producer. He or she has heard it all before, so make sure you have your facts straight. Keep it simple: Tell why you feel your appearance will benefit their audience, and that you'll do everything you can to help it go smoothly.

Several cities have local publicity directories that list radio and TV talk shows, the producers and hosts, what they are looking for, and how to contact them. Check with your local advertising and publicity club. In Los Angeles, for example, the *Publicity Club of Los Angeles Directory* is an annual and costs $105. It's worth it.

Press Conferences. In order to determine whether to do a press conference, use the following checklist. If you don't answer yes to at least three of the questions, you don't need one.

1. Is this a *major* product or company announcement?
2. Will this have a *major impact* on the audience?
3. Is this *so complex* that it can't be explained in a release?
4. Do you have a celebrity or politician involved?

PUBLICITY TOOLS

There are many tools you need to know how to use. We will discuss each one, and how and where they fit into the scheme of things. You'll find several examples of forms for these tools in the Appendix at the end of the book.

1. Media advisory
2. News release
3. Public service announcement (PSA)
4. Company backgrounder
5. Fact sheet
6. Op-ed piece
7. Letters to the editor
8. Feature articles
9. Self-syndicated articles
10. Photographs
11. Question-and-answer formats
12. Clip sheet—print
13. Clip sheet—electronic
14. Media analysis grid
15. Media tracking sheet

Media Advisory

The media advisory is an advance indicator that you intend to stage some sort of press event, conference, or publicity promotion sometime in the near future. It gives the basic information, and outlines in paragraph form why the media should cover the event.

News Release

This is the core element of your publicity campaign. From this will flow articles, feature stories, and other documents, either written by you or by members of the media. (It is no longer appropriate to call this a press release, because the media consider the word *press* to mean "print." The broadcast folks don't take kindly to being excluded, even unintentionally. So call it a news release.)

The release gives the basic facts in journalistic format: who, what, where, when, why, and how. It is written in inverted pyramid style (see p. 159), and kept very simple. It's best to limit it to one double-spaced page. Send your release at least one week in advance of the day you want it printed. For a complete discussion of news releases, see the news release section later in this chapter.

Public Service Announcement

The public service announcement, or PSA, is normally used by nonprofit organizations and charities. It is sent to both television and radio stations. It is sent two weeks in advance of the event to the public service director or public affairs director.

For radio: send the PSA along with a cover letter. For television: send the PSA, cover letter, and a slide with your name, logo, and phone number. Remember, television is a visual medium. What sense is there in getting a PSA on the air with only the CBS eye staring the viewer in the face?

Company Backgrounder

The company backgrounder gives information on the principal player in your business—namely, you. If more than one person is involved, use the format shown on the form in the Appendix, briefly describing each of the key players.

Fact Sheet

For inventors, the fact sheet is a one-page, double-spaced spec sheet on the invention. While you will be sending this with a news release, it is possible that editors might miss key facts. The fact sheet fills in any important gaps and it may save an otherwise imperfect release.

Op-Ed (Opposite Editorial) Piece

This is a more sophisticated tool, but one that can give you tremendous exposure. It is, in effect, a guest column, run opposite the editorial page. It may or may not relate to previous opinions printed in the publication.

Most op-ed pieces are written by syndicated columnists, so your chances for space are small. However, with a well-written piece and good credentials (the press currently seems to love women and academicians with Master's degrees or above), you should be able to get one in.

Letters to the Editor

This is a relatively easy way to get into the press. Most publications offer their readers a chance to "sound off" about one thing or another. Make sure you keep your letter brief (about 150 words), direct, and to the point. Don't ramble. Tightly edit your writing. And be sure to include your name, company name, and city at the bottom. Too many letters show name and city, but no company name. What a waste of good PR space!

Feature Articles

Here's where the smart inventor stands out from the crowd. Develop a 750-word article on how your invention is going to save humankind something: time, money, natural resources, for example. Editors love timely causes; the latest are environmental, economic, and social. Stick with those three, tie in how your invention solves a problem in one of those areas, and you should have a winner.

The feature is usually a one-shot deal. You can either self-syndicate this article to various publications, or offer it to the editor of your local paper.

Self-Syndicated Articles

These are feature articles you write on a regular monthly or semimonthly basis and send in to association newsletters, journals, weekly papers, and perhaps even trade magazines. (The latter will require intense negotiation with the editors.)

Print them out double-spaced, one side of the paper, on plain white bond. Enclose a cover note, and include a SASE (self-addressed stamped envelope). Make sure you track who you send them to, and which articles are sent.

Offer to write a regular column on different inventions and how they influence the community. (In the later case, you are the resident expert on inventions in general and cannot push your particular product more than once.)

Photographs

Photographs can be used to supplement your release, or they can be sent by themselves with caption sheets (these are called stand-alone photos). Whenever you send a release discussing your invention, send an action photo along with it. You can also send stand-alones to the new product editors of trade magazines.

The caption sheet should be printed out on plain white bond, double-spaced, at the *bottom* of the sheet. Attach the top portion to the back of the photo with clear tape. (Do not use paper clips—they can ruin photos.) Do not write on the back of the photo, or put labels or anything else there; they might bleed or show through, and the publications will not be able to use the photo. Fold the caption sheet in half, and bring it up over the front of the photo. When the editor opens it up, the bottom caption half falls open, and he or she can read it.

Question-and-Answer Formats

These are normally for talk shows. Whenever I do a talk show, I ask the producer if it would be okay to send, in advance, a list of the twelve most commonly asked questions about my seminars. They almost always agree.

What you are doing is stacking the deck. Send them twelve good questions that cover *everything* important about you, your company, and your invention. Since you'll be able to rattle off the answers, you look good and become the instant accessible expert. (Be sure to bring a copy of the questions with you to the interview.)

Clip Sheet—Print

Clip sheets are copies of articles, releases, etc. that have already been published. Take the three most recent articles printed about you or your invention. Reproduce them on your letterhead, on one side, one sheet, along with the folio of the publication at the top, and send them with your release. (The folio is the portion at the top or bottom of the page of a publication giving its name, day, and date. For example, *Los Angeles Times,* June 12, 1987. Don't type this; make sure it's from the paper.)

If the articles are too long, reproduce just the paragraph before and after you are mentioned, and circle your name. In all cases, make sure the headline is also reproduced.

Keys. You should have a variety of clip sheets in your files. Be careful who you send them to. Have one clip sheet in a folder labeled "Local Weekly Clip Sheets," another labeled "Trade Clip Sheets," and a third labeled "Dailies Clip Sheets."

The best method is to send clip sheets on a peer-to-peer basis. (This is what you need *Gale's Directory of Publications* for; the circulation figures will help you decide the peer relationships.) The *New York Times* or *Wall Street Journal* will not be impressed if you've been printed in the *Podunk, Iowa Dispatch.*

Finally, don't send clip sheets to direct competitors (within 100 miles of each other). They probably won't print the information. Every editor likes to feel he has gotten a "scoop."

Clip Sheet—Electronic

While it's not possible to send copies of broadcast news and interviews, you can send a list of the stations and dates of your interviews. On your letterhead, list the station, date you appeared, name of program, producer, host, and city and state.

Media Analysis Grid

This is used to determine what to send, and when. Remember, you always want to leverage your chances for success.

Let's say you've picked twelve trade magazines to send information to. What should you send first, second, third? (A media campaign is not just a release; it includes follow-up calls, articles, opinion pieces—everything we've mentioned in this section.)

Look at the information from your contact card on what publicity materials they take. Using the grid (see the Appendix), list one magazine per box at the top. Check which materials they take. Do this for each magazine.

When finished, you will have some who take releases (everyone does), some who use articles, some who use photos, and so on. If twelve take releases, seven take articles, five take op-ed pieces, etc., that's the order of publicity tools you use. First send releases, next write an article, sending it to the seven who might use it. Then do an op-ed piece, sending it to the five.

Publicity is a numbers game. This grid will help you leverage the odds in your favor.

Media Tracking Sheet

Whenever you send an information tool out, you must track it. Of course, you'll note this on the contact card. In addition, for each type of tool sent out, use a separate media tracking sheet.

Using the above example in the media matrix section, have one tracking sheet for the releases, one for the articles, etc. This method gives you a picture of how you're doing.

You'll know within three weeks of the publication date whether or not a certain media has used your story. If you haven't received a tear sheet, or copy of your article, by then, they are not going to use it. By using a yellow highlighter to cross out the line of that particular medium, you will still be able to see where you sent the information to, as well as what worked versus what didn't work.

THE PRESS KIT

The ingredients of a simple press kit are:

1. Media advisory (optional, use if an event is taking place)
2. News release
3. Fact sheet
4. Company backgrounder
5. Product photograph (action shot)
6. Clip sheet
7. Clip sheet—electronic
8. Binder or cover

When to Use the Press Kit

The press kit is most often used at events, conferences, and trade shows. When contacting the media the first time, send only a cover letter, action photo, release, and product fact sheet. Too much is overkill. Hold back on the full press kit until they ask for more.

THE ALL-IMPORTANT NEWS RELEASE

The secret to finding ideas that will sell your releases or stories to the media is to look around you at work or at home. Writing about what's familiar makes writing comfortable, and when you're comfortable, you create and write better.

What does your invention do that will make people's lives better? How will it change our world? Why is it better than others like it? These marketing questions from a previous chapter will help you develop an angle on which to base your release. (For further information, see my booklet *101 Ideas for News Releases*, P.O. Box 2667, Chino, CA 91708-2667.)

Your invention must tell a story that is local, unique or unusual, timely; that concerns people; and that creates human interest. If any one of these factors is missing, it is not as strong as it could be. You can get publicity with one or two missing, but put in all five criteria, and you've got a solid story.

The Rules for Releases

The news release will be the most important tool you use. If you master how to write it, where to send it, and who to send it to, you should get more than 50% published or broadcast. There are two rules to help you write your news release: the inverted pyramid style and the five Ws and H formula. (See the Appendix for format.)

Inverted Pyramid Style. This rule says to put the *most important* point of your release in the first paragraph, the second most important point in the second paragraph, and the third in the third. (Never have more than three selling points in one release. Otherwise, with too many mixed messages, the reader may get confused).

How do you know which is the most important point? Simple! Make a list of six key selling points about your invention. Rank them in order, from the most important (#1) to the least important (#6). Then #1 goes in paragraph 1, #2 in 2, etc.

The Five Ws and H Formula. The five Ws and H stand for: who, what, where, when, why, and (sometimes) how. In other words, as you write your release, answer the following questions about your invention:

- *Who* are you? *Who* is your company?
- *What* is your invention?
- *Where* can it be used?
- *When* will it be available?
- *Why* is it unique?
- *How* does it work?

Answer all these key questions, put them in the first paragraph, and you have a good opening for your release. By putting all the key facts in paragraph 1, if the editor takes your release verbatim and still has to cut, the main facts get in. (Editors cut from the bottom up.)

Another reason for this formula is that you must answer all the possible questions editors might have about your release. Editors are busy people. They don't have time to call if there is missing information; they'll just toss the release. Use this formula, and you won't leave anything out.

Not only must the release answer the questions; the answers must be complete. For example, if you had a sentence that read "This invention will be displayed at a press conference on Sunday at 2:00 P.M.," you'd be in trouble. There's a hole in it. Which Sunday, what's the date, and where will it be held? *Missing facts kill a release!*

Notice these other key elements in the format:

1. Double-space your release.
2. Put the contact name and number at the top.
3. Only four or five short paragraphs per release.
4. No longer than two pages, preferably one.
5. Use a brief, powerful headline.
6. Use "-End-" or ### to indicate the finish.
7. Write in simple, eighth-grade-level style.

IN SUMMARY

You've received a lot of information on publicity in this chapter, far more than you can absorb in one sitting.

Reread this chapter several times, using the forms in the Appendix as guides. Then try an experimental release or article, and show it to several friends. The more you write, the better you'll do.

13

Advertising: Shouting Your Message Out Loud

I have good news for you: Advertising is both relatively easy to do, and it *can be* less expensive than you think. The key is to find the right media and to give your customers the right message at the right time. You might think that's obvious but very tough to do. It's not. The techniques in this chapter will show you how.

THE VARIOUS WAYS TO ADVERTISE

Without advertising, nothing happens. If the world didn't advertise daily, very few products—except staple items, like soap and food—would be sold. (Of course, they advertise, too, in order to persuade you to buy certain brands.)

Selling makes the world go 'round. Without the sale of products, there would be no manufacturing. And you, as an inventor, would be out of business. *You need advertising, and you need to advertise!*

The challenge is choosing the right vehicles and being able to pay for their use. There are literally hundreds of ways to advertise your

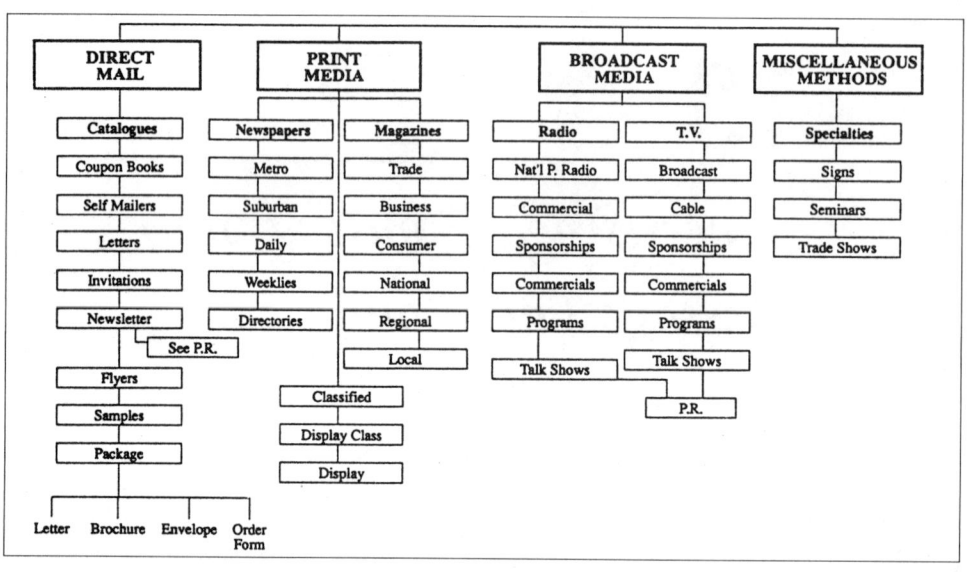

FIGURE 13.1 Advertising Methods

invention. Figure 13.1 shows the most common methods, and where each subcategory fits.

As an inventor, you will be concerned with four main advertising categories: direct mail, print media, broadcast media, and what I call miscellaneous methods. We will explore all four areas.

DIRECT MAIL

Direct-mail methods of advertising include catalogs, coupon books, self-mailers, letters, invitations, newsletters, flyers, samples, and direct-mail packages.

Catalogs

Catalogs are useful only if you have multiple products to sell (at least two dozen) and have a defined, targeted audience to send them to. Catalogs are very expensive to produce, and they are not normally used by small business owners or inventors.

Coupon Books

Coupon books are excellent for retailers. But again, unless you have multiple products, coupon books do not lend themselves to invention marketing in the normal sense.

Self-mailers

Self-mailers can be very useful to an inventor. In a self-mailer, you type the prospect's name one time, and it reproduces on an invoice, statement, or short ad letter, as well as on the outer envelope and the inner return envelope. Self-mailers are expensive and should not be tried by the novice.

Sales Letters

Sales letters are the *premier* way to market your invention, within the realm of direct mail. A personalized, brief, on-target sales letter can deliver more punch than a prizefighter. You should immediately learn the necessary techniques in preparing one. (See the section later in this chapter).

Invitations

Invitations are a good, inexpensive way to promote your business. Invitations can be sent at the birth of your invention and at any time you plan a major promotion. In the latter case, holding a party to show off your new invention is the perfect place for using an invitation. Announcements can be used as post-promotion, telling the media and your prospects what happened at that *fabulous* party.

Newsletters

Newsletters are often sent as part of a full direct-mail package or as stand-alones. Newsletters are also part of public relations, in that they have two distinct yet interwoven functions: Inform the receivers (the publicity aspect), and advertise the invention(s) for sale.

Flyers, Circulars, and Brochures

These are inexpensive vehicles that should be part of your initial marketing efforts. In fact, it is *essential* that you have a brochure describing your invention. But avoid putting *everything* you do in the brochure. You can't be all things to all people. Likewise, don't sell everything in one brochure. Too many messages confuse the reader.

Be sure to have good black-and-white photographs (action shots) taken of your invention for reproduction in the brochure. Flyers, circulars, and brochures can be incorporated into a full direct-mail package, or they can be stand-alones.

Samples

If you have samples of your product and they are not too bulky or costly to make and ship, send them along with a sales letter and brochure to your prospects. This makes an excellent direct-mail package. (Service businesses can also use sampling, in the form of *free* consultations, free estimates, etc.)

Direct-Mail Packages

The full direct-mail package is an essential advertising vehicle for the serious inventor. It consists of a sales letter, a return envelope, a response card or form, and a brochure. A full discussion of the direct-mail package follows later in the chapter.

PRINT MEDIA

Print media advertising can be found in a number of different vehicles. The most commonly used are newspapers and magazines. Newspapers are classified into metro (metropolitan) editions, suburban papers, dailies, and weeklies. I also include specialty directories in this category. Magazines include trade, business, and consumer publications. They fall into subclasses of national, regional, and local.

Newspapers

Newspapers are classified according to their publishing schedules. For purposes of this discussion, metro editions are usually large-city daily newspapers, with circulation figures in the millions. Examples are the *Los Angeles Times, Boston Globe, Chicago Tribune,* and *New York Times.*

Suburban Papers. Suburban newspapers are published either weekly, biweekly, or triweekly. What makes a newspaper suburban is its town's proximity to a large metropolitan area.

Dailies. I classify dailies as papers having under 1 million circulation, not located within metropolitan areas—for example, the *Orange Coast Daily Pilot* in Costa Mesa, CA. While definitely a daily, it could also be classified as a suburban paper because it's so much smaller than the other major daily serving the area, the *Orange County Register.*

Weeklies. Weekly newspapers are published once or even twice a week, have a limited circulation, and are usually located on the fringe of a metro area. Examples would be the Highlander chain in southern California. Each community edition is no more than 12,000 copies. Yet with fifteen editions, it's a force not wise to ignore.

To find out where certain publications lie, check with *Standard Rate and Data Service.* Most ad people use their directories as a classification system.

Specialty Directories

These are publications that come out annually or semiannually, and are usually sold through charities or chambers of commerce. Examples are the Boy Scout or Girl Scout directories, church bulletins, etc.

Community relations are all very well, but stop to ask yourself, "Can this really help promote my business or invention?" If the answer is yes, you might *possibly* think of investing some money in an ad, as long as it's discretionary money. If not, don't do it. Basically what you're doing in this case is giving to a charity.

I know there are those of you who feel charity is necessary, and we must support needy organizations. I couldn't agree with you more—but *not* with your promotion dollars! In this case, charity begins at home.

BROADCAST MEDIA

The broadcast media consist of radio and television, both of which include commercial, public, and cable stations. You can use *SRDS* volumes on radio and television for advertising rates and information. However, since *SRDS* only prints what the stations send them, I suggest you start with *Broadcasting/Cablecasting Yearbook*.

Radio

Radio advertising includes the use of public and commercial radio stations. Vehicles are sponsorships, segments, commercials or spot buys, program brokering, and talk shows.

You cannot advertise, per se, on public radio or television. You can, however, buy sponsorships from your local PBS affiliate. The PBS system is supposedly commercial-free, although that may change in the next decade, as they continue to lose funding.

Commercial radio stations are licensed by the FCC and allowed to air commercials. The problem is which kind of commercial ad to buy and how many. I will explain the process to you later in the chapter.

Sponsorships. These can be either total program sponsorships or segment sponsorships. In the entire program, you buy *all* the airtime available, and only your commercials run. This approach was common in the golden days of television and radio, when the cigarette companies would sponsor entire shows.

The only half-hour programs now in radio are business-oriented. You might consider a sponsorship, if the price is not too high and the audience is right for your message.

Segments. Sponsoring segments are easier and cheaper. A segment is a news or sports segment, once a day in the morning or evening commute hours. Although segments can be expensive, they do generate listener response quite well. However, this is a more sophisticated form of radio advertising. You should wait until you have mastered the 30- and 60-second spot commercials.

Commercials or Spot Buys. Actual radio commercials and/or spot buys are the form of radio advertising purchased most often. The aura surrounding radio stations makes most people think it's too sophisticated or expensive a medium to use. (See later in the chapter for a full discussion.)

Program Brokering. Find a local radio station with a good-sized listening audience, but not the largest. It shouldn't be a major station.

Offer to buy a half-hour segment to explore inventing problems, including a discussion of how your business solves them. Bring on a few other inventors and talk about your businesses. If you can get the others to share a portion of the expense, you are program brokering. Some stations will not allow this practice, so check with each station in your city.

When I was living in San Diego, I did two radio programs in just this fashion. I would buy a half-hour time period for thirteen weeks and resell sponsorships, segments, and spots.

Talk Shows. Being on a talk show is tremendous publicity. If you want further promotion, you might consider buying a few spots around your interview. But avoid overkill. Remember, publicity has greater credibility than advertising.

Television

The various types of television advertising methods on broadcast or cable stations are sponsorships, commercials, programs, and talk shows. Most of what I said about radio commercials, spot buys, and talk shows applies to television as well. The big difference is cable TV.

Cable Television. Cable TV is a relatively new phenomenon, having been introduced within the past 30 years. It was started because of poor signal reception due to mountains. Once people understood that cable dramatically increased signal reception everywhere and made for better television, everyone jumped on the bandwagon. Carrier companies were formed.

Within each community in the United States, different cable companies are licensed to lay cable and feed broadcasts through the system. At present, over 90% of the country is now hooked up with cable. Each community cable company is charged with the responsibility for providing both regular programming and local origination channel (LOC) programming.

LOC Programming. LOC programming usually involves city council meetings, church plays, community theater, and the like. However, the enterprising inventor can make great use of this vehicle for advertising.

FCC rules state that every cable carrier *must* carry a certain amount of local programming. As a local community small business owner, you can make sure they do their job. Call the carrier and ask for the person in charge of the local programming. Ask if they have a local talk show. If they do, ask how to get on.

If they don't, start your own—either an inventor's workshop or a regular talk show format. Ask the staff to help you. They're obligated to do so, and if they hesitate, insist on seeing a copy of the LOC rules.

You can also advertise on cable channel crawl screens, although I don't recommend it. No one stays on the channel when it just shows a moving screen with ads and background music.

The one exception I know of is Mammoth Lakes, CA. During the winter skiing season, everyone watches in the morning for the weather report. At the top of the screen, commercials in the form of moving print ads constantly scroll across. I don't think they do much good, but people I know say they get business from them.

MISCELLANEOUS METHODS

As stated at the beginning of the chapter, there are numerous ways to advertise your service or product. Many of these will not be right for you, because the image and message of the particular media vehicle will not coincide with those of an inventor. However, there are several that are compatible: advertising specialties, signs, seminars, trade shows, and telemarketing.

Advertising Specialties

These are commonly known as giveaways or incentives. Examples would be pens and pencils, paper cubes, and keychains that have your company name, slogan, or imprint on them. Small businesses use these specialties as an advertising vehicle.

Signs

While most people equate signs with storefronts, different types of signs can be very beneficial to the inventor. The signs on your packaging, for example, tell at a glance what the product is and whether it's appealing to potential buyers.

Point-of-purchase (POP) signs are perfect for store and trade show displays of your invention. Large window posters, and handbills, may definitely be needed if you sell to a store.

Outdoor signs, billboards, signs on benches at bus stops, balloons, skywriting—all of these are far outside the realm of any budget you would initially have. Leave these types of advertising methods to the major players.

Seminars

Seminars are an excellent vehicle for the inventor, especially if you have a technical invention that needs a lot of explanation. These educational workshops can really be super selling opportunities for you, provided you handle the audience with dignity and keep your "sales pitch" low-key. Since much of what you tell your audience will be in the realm of education, seminars are also good publicity vehicles, for both pre-event and post-event press coverage.

Trade Shows

Trade shows are so important that I've devoted an entire chapter to the processes involved, Chapter 14. Suffice it to say trade shows can provide abundant opportunities for you as an inventor, from a public-relations standpoint as well as that of an advertising and sales tool.

Telemarketing

Telemarketing is simply "dialing for dollars." With the cost of a field sales call becoming more expensive these days, more and more companies are using the phone for prospecting and selling. An excellent book for learning telemarketing is Gary Goodman's *You Can Sell Anything by Telephone* (Prentice-Hall, N.J., 1984).

Yellow Pages

While good for the retailer or service businessperson, the yellow pages are not right for inventors. The cost can be prohibitive, because you must contract for an entire year.

THE DIRECT-MAIL PACKAGE

A direct-mail package has five basic components: a brochure, a sales letter, an order form, a return envelope, and an outer mailing envelope. Exclude any one of them, and you'll have an incomplete, ineffective direct-mail package.

The Brochure

There are many ways to design brochures. Two of the most common are the six-panel (8.5 × 11-inch paper, scored and folded twice), and the eight-panel (8.5 × 14-inch paper, scored and folded three times). They're made from high-quality paper and usually at least two colors (black plus one other).

The basic design of a six-panel brochure is not complicated if you follow a format. The format is based on the five Ws and H formula you learned in Chapter 12: who, what, where, when, why, and how.

In a brochure, you must tell your story as briefly and completely as possible. Using the five Ws and H formula makes this easy. Just answer the questions: Who is the company? What is the product or service all about? Where is it available? When is it available? Why is it a benefit to the reader to buy it? (*the most important point of all*) How much does it cost? To lay out a six-panel brochure, follow this simple format, shown in Figure 13.2.

- *Panel 1.* Front cover: Headline, logo, name.
- *Panel 2.* Inside front: Who, what, where questions.
- *Panel 3.* Middle inside: When, why, how questions.
- *Panel 4.* Inside back: Order form or action step.
- *Panel 5.* Outside back: Return postage/mailing information.
- *Panel 6.* Opposite middle inside: Testimonials.

This is the format I use, and it always seems to work out fine. It's crisp, concise, and to the point. I make sure to answer all the questions, and I sometimes even put the words *who, what, where, when, why,* and *how* in bold caps with bullets for emphasis, as shown in the figure.

The Sales Letter

The sales letter is the most important part of a direct-mail package. It can, however, stand alone, and it is used quite extensively in business. No matter where you use it, there are basic rules to follow.

1. *Give it that personal touch.* Nothing helps sell a prospect better than that homey, warm, personal feeling. It's as if you wrote the letter to a family friend or relative. You may not have the time or money to customize each letter. But if you write in such a way that it seems like you did, you stand a good chance of success.

2. *Do multiple mailing campaigns in stages.* All sales professionals know it takes seven contacts to make a sale. By sending multiple sales letters, you leverage your chances for a positive response.

Panel 1

PLAN AND PROTECT YOUR FUTURE

A seminar for active and retired classified employees

Saturday, April 1, 1989
8:30 am to 1:30 pm

Sponsored by California School Employees Association

Panel 6

PLEASE ATTEND

As president of the Board of the Public Employees Retirement System and past state president of CSEA, I can't emphasize enough how important it is for classified employees to plan ahead for retirement. Once retired, it's equally important to keep abreast of changes in the laws that affect you and programs that can help you get the most from your retirement dollars. I plan to attend this important seminar and hope you will, too.

Bill Ellis

Your Retirement Committee has worked hard to plan this program for you. We urge employees of all ages to learn from the experts. We also encourage all members who are retiring to become a retired member and benefit from current and future CSEA programs.

Muriel Lightfoot

Panel 5

WHY?

To help all classified employees understand the need to prepare for the future, and to bring active and retired members up to date on information to enjoy a healthy and happy retirement.

Get information on the following programs:
- Bronson Vitamins (new-sponsored mail order vitamin program)
- CSEA Special Services
- E.A.R.S.
- EJS Insurance (Medicare supplement, long-term care, prescription drugs and group dental program)
- Hypertention Control Council
- PERS
- Retirement Committee
- Zahorik Company (tax shelter annuities)

DOOR PRIZES AND MORE

Among the many door prizes is a three-day, two-night stay at the beautiful London Bridge Resort in Lake Havasu City.

Get a free hearing test, compliments of E.A.R.S. during lunch and following the program.

BUSINESS REPLY MAIL
FIRST CLASS   PERMIT NO. 5432   SAN JOSE, CA

POSTAGE WILL BE PAID BY

CALIFORNIA SCHOOL EMPLOYEES ASSOCIATION
Orange Field Office
326 West Katella Avenue, Suite E
Orange, CA 92667-9964

NO POSTAGE NECESSARY IF MAILED IN THE UNITED STATES

Panel 2

➡ WHAT?

A seminar packed full of information about financial preparations, health and travel to help you plan for, and enjoy, your retirement. An important update on Medicare and how Medicare supplements and long-term care plans work are also included. Free blood pressure and hearing testing, lunch, prizes.

➡ WHERE?

Maruko Hotel
295 North East Street
San Bernardino

From Freeway 215 take 2nd Street exit, go east on 2nd Street to E Street, turn left to Maruko Hotel, turn right into parking lot.

➡ WHEN?

Saturday, April 1, 1989
8:30 am to 1:30 pm
(registration 8:30 am to 9:00 am)

➡ HOW MUCH?

Absolutely free!
Complimentary coffee, pastries and light buffet luncheon

➡ WHO'S DOING IT?

The California School
Employees Association

Panel 3

PROGRAM

Financial Readiness

Financial preparation before and during retirement is critical. Attorney Richard Barr will describe simple ways to plan your financial future. He will also provide retirees necessary information on tax laws and wills. Mr. Barr has been an attorney for sixteen years and has had a general law practice in Santa Ana specializing in estate planning and preparation of wills.

New Medicare Laws

New Medicare laws took affect January 1, 1989. A speaker from the Health Care, Financing Administration in San Francisco will explain exactly what the changes are, how you will benefit and what it will cost you.

Hypertension Control

Jose Marquez, director of the Inland Empire Hypertension Control Council, will describe signs of hypertension and suggest ways to avoid becoming a victim of a heart attack or stroke. Mr. Marquez will provide a staff of volunteers to give free blood pressure test during the lunch break and at the close of the seminar.

Travel Wise

Simple tips on travel planning can help make your travel dreams come true and make all your trips both economical and enjoyable. Annette Passaretti, president of Mature Travelers, Inc. will give money-saving ideas for retirees, tell you how they work and how to get the most from your travel dollar.

Medicare and Supplements and Long-term Care

The new CSEA-sponsored programs are designed especially for our retired members. Each program may save you money depending on your individual needs. Susan Goff of EJS Insurance will give a brief overview of the coverages and be available to answer individual questions.

Panel 4

Please return by March 24.

YES! I will attend the Retirement Planning Seminar

Name _____
Address _____
City, zip, state _____
Phone (home) _____ (work) _____
School or community college district _____

FIGURE 13.2 A Six-Panel Direct-Mail Brochure

3. *Put the "you" in everything you do!* That's a great slogan, and one that has real merit. (The personal touch is most important, yet large corporations think it means adding a person's name by computer. You and I can tell just by looking at the typefaces that the letter has been computer-generated.) If you truly customize your letters and put the "you" (implied or actual) into the copy, you'll get the sale. For example, "Here are six reasons why you will save money with our invention." Saving money is always fun, and the six reasons arouse curiosity. The "you" is even stated in the sentence. Don't be shy. You can't afford to squeak like a mouse in your sales letters when others are roaring like lions.

4. *Make sure your writing skills are sufficient.* Know your grammar and punctuation. If you're rusty, get help from a freelance writer or editor, or read up on the subject. Write to the level of sophistication (or lack thereof) of your target audience. You will know what that level is based on your previous market research.

5. *Make sure your letters and envelopes look professional.* Invest in good stationery. Don't use computer labels, even transparent ones—they imply that you don't care. Instead, print out the address on each outer envelope. Make sure your printer has letter-quality type. Use postage stamps. Postage meters imply volume solicitations and a lack of care.

6. *Use the AIDA formula for writing.* AIDA stands for attention, interest, desire, and action. You must (a) get their attention, (b) arouse their interest, (c) create a burning desire for your invention, and (d) ask for the order (action step).

AIDA Defined. The *attention* step should be a headline in a small box at the top center, discussing the key benefit.

Interest and *desire* are found in the body copy. Here's where you capitalize on two or three top selling points, written from a benefit-laden point of view. (Benefits are what your invention does for your customer, not what you do.)

The *action* step calls for the order. You might say something like "Call Now for Further Info." This action step tells the prospect to do something *now!*

Use only two or three key selling points in any one letter. As a matter of fact, the same principle applies to all advertising tools: brochures, ads, news releases, etc. Any more than three, and you have too many messages flooding the reader. It won't work.

How do you find the top three? Easy. List the key sales points about your invention. Then rank them, and put the top three in your sales letter.

7. *Write at a sixth-grade reading level.* This will not insult them. Did you know the number-one magazine of all time, the one with the highest circulation, is *Reader's Digest*? Why? Because they write the way people read: simply. The average person reads at sixth-grade level. If your invention is technical, go for a tenth-grade comprehension level, but no further. The higher the level, the fewer people will be able to understand and buy your product.

How do you determine writing level? If you have a computer, buy *Right Writer* or *Gramattik VI*, which are powerful software programs that analyze your copy. They not only tell you the level of your writing, they also show you ways to improve. For $69, they're a steal! No computer? Get one!

8. *Write the way you talk.* I tell all my students that writing is nothing more than verbal expression on paper. When we talk to our friends and relatives, we're at ease and tend to speak simply. Yet when we write, we become flowery. Write as if you were talking to a friend.

9. *Keep it brief.* Contrary to popular opinion, short letters result in more sales. I recommend a concise, one-page sales letter. Who has time to read more than that? Long letters get ignored. Keep yours short.

Benefits vs. Features. Features are the specifications of your invention, such as size, shape, weight, materials, etc. Benefits are how the specifications directly support your potential customers—in other words, "What's in it for me?" Remember to put the most important benefit first.

For example, a feature of your invention might be that it's light in weight. The benefit to your customers is that lifting it will not cause back strain or aching muscles.

In summary, the successful sales letter promises benefits, tells the prospect what he or she will get, gives examples to show the benefits work, summarizes the benefits, and asks for the order.

The Order Form

The order form, the third element in a direct-mail package, can be a self-mailer, a response card, a tear-off sheet or card, or the return envelope itself. The order form is the second most important aspect of the package. While potential customers may read your appeal in the sales letter, if your order form is incomplete or inefficiently designed, they won't place an order. So design your form ASAP.

A good order form has the following characteristics:

1. It contains room for all the necessary specifications (size, price, etc.).
2. The specs must be in the appropriate sequence.
3. The form must be designed to sell the product.
4. The form must be easy to understand.
5. It must be simple in design.
6. It must be easy to fill out and process. (Give the customer enough room to write!)

Only if these six criteria are met can you be sure you have an order form that will make it easy for the buyer to buy. And after all, that's the name of the game—make it easy for the buyer to buy. (It's so important, I've repeated it twice in one paragraph.)

The example on the facing page shows a good, simple, clean order form for a mail-order book. Notice how the wording gets immediately to "you."

Self-Mailer. The concept of a self-mailer is similar to the one-write system of record keeping. The outside mailing envelope, inside sales letter, order form, and return envelope are all constructed in such a way that when you address the outside envelope, you're writing on all the elements at the same time.

A self-mailer is normally used for statements of accounts, invoices, and due bills. However, with a little creativity and imagination, it could work as part of a direct-mail package for your invention.

YES, I WANT TO ORDER *HOW TO SELL AND PROMOTE YOUR IDEA, PROJECT, OR INVENTION!*, Reece Franklin's powerful 200 page, 8½″ × 11″ workbook, at $19.95 each. I'm excited, and I can't wait! I understand I will receive my copy by return mail within 3 weeks. If for any reason I'm not satisfied, I can return the book, in resellable condition, for a prompt refund.

I've enclosed my check for _____ books @19.95 each plus $2.00 shipping and handling. Total amount enclosed is: $_____ .

Here's my charge card number: _____

Master Card____ Visa____ American Express ____ Exp. Date _____

Signature of Card Holder _____

Name _____

Company _____ Title _____

Address _____

City _____ ST _____ Zip _____

Business Phone ( ) _____ Home Phone ( ) _____

Return with check or charge card info to: Prima Publishing, Box 1234RF, Rocklin, CA 95677. Telephone (916) 786-0426

Response Card. This is a separate card that can be printed on both sides or on just one side. If printed on one side, it is inserted into a return envelope. If printed on both sides, the back is usually a business-reply return mailer. This version is not very fancy, but it can be less expensive than using an additional return envelope. The one problem is that the ordering information is visible to anyone who picks up the card, and many people prefer keeping such information private. Consider all factors before printing.

The response card inserted in an envelope is the most common way of designing an order form. And most sophisticated direct mailers use a self-stick label to get the customer involved. You know the one I mean, the Ed McMahon special: "Just place the YES sticker on the card, and we'll send your free gift and subscription right away!"

Tear Off Sheet or Card. This is an order form at the bottom of a sales letter, perforated for easy removal. It works best with a single-page sales letter. Using this tear-off form gets the reader involved in the buying process physically, which is a positive action step. (Remember AIDA? A = Ask for the order.)

The tear-off sheet, like a coupon response form, should be at least as big as a dollar bill. The headline must stand out, and there must be enough room for the customer to write.

The Return Envelope

Sometimes return envelopes can do double duty. By adding a tear-off flap which is the order form, you can incorporate two pieces of paper into one, initially. Many department stores with revolving charge-account customers use these to increase the participation of the bill payer.

In this case, what's printed on the flap of the return envelope (which is used to send the payment) is a product that can be ordered right then, just by tearing off and inserting the flap. This is called piggybacking, and it's very effective.

The Outer Mailing Envelope

The final essential component of your direct-mail package is the outer mailing envelope. While we're discussing it last, it is the *first* thing the prospect or customer looks at, so it should be done correctly.

The idea is to get the prospect or customer to open the envelope and look inside at all the goodies. You can't expect them to be excited if the envelope telegraphs "sales message."

What about the Ed McMahon type, with contest and other messages written all over the outside? They send out millions and can afford to have thousands thrown out, knowing they'll get their one-half of 1% response, and this will equal thousands. You can't afford to send that many.

So what do you do? Make your mailing envelope look like a personal letter to your target audience. Use a regular stamp, not a postage meter. Hand-write the address, for a more personal touch. Don't put your company name as a return address; use your own name.

The more you make your audience think the package is coming from a friend, the more people will open it.

Keys to Direct-Mail Success

When to Send a Direct-Mail Package. Some believe that the best time to send business mail is March through May. But I believe you can send direct mail every month of the year and receive some kind of response. It all depends on the invention, the season, and your offer. There are too many variables to say any time is better than another.

However, there *are* buying cycles. You need to understand the cycle of your industry. This information is available to you through your industry association.

Expected Response. The rule of thumb is that 0.5–1% of your list will respond to your offer. Of course, all rules can be broken, and the variables often come into play. Some of these variables deal with the buyer's perception; some deal with the nuts and bolts:

- If you have a bad list, your response will be down.
- If you have a bad offer, your response will be down.
- If you have a bad product or invention, no one will buy it.
- If your product or invention is too expensive, no one will buy it.
- If you fail to plan, you will have a poor showing.
- If you have a highly targeted audience, your response should be higher.

Basic Rules. There are seven keys to success in direct mail, according to advertising guru Jim Lorenzen. (Lorenzen, Jim, *Ad Strategy That Works*, FL: Lorenzen Associates, 1983.) These are:

1. *Know what you want the mailing to do.* The mailing *must* return at least 0.5% response if it is to be considered successful. But you must know what type of response you want. Do you want just leads for follow-up, or do you want product sales? Whatever your desired results, you must structure your offer accordingly.

2. *Write the copy so buyers understand the benefits.* Again, I stress the importance of benefit-laden copy, not feature-laden. You *must* answer the buyer's question, "What's in it for me?" If you don't, you'll lose the sale. The copy must cover all the facts: benefits, sizes, colors, shapes, composition, cost, differences from other products. In short, satisfy the reader.

3. *Layout and format must tie in with the overall plan.* In your marketing plan, you have overall goals and objectives you want to meet. Out of these come an image of the way you want your company to be perceived. If your layout and format are giving off the wrong signals and the rest of your advertising and marketing deliver another message, the buyer will become confused. And confused buyers buy elsewhere.

4. *The list is the key.* If I were to give you one key element in direct mail, it would be the importance of list selection. If you have a customer list already, great! You know their hot buttons and the types of offers they want. If you do not, find a list that matches your potential customer profile (developed in the chapter on marketing, Chapter 11).

List selection can be very complicated and time-consuming. I work with list brokers, who do the legwork for me. For a selection of list brokers, use the *Standard Rate and Data Service* volume on direct mail.

Under each industry category, you will find at least a dozen brokers, with information on prices, how they offer their lists (mag tape, Cheshire or peel-off labels, diskette), if a sample mailing piece is required, and how long it takes to receive the list.

The number-one question to ask the list broker is, "Where did you compile your list from?" If he is simply using other people's lists or compiling from the phone directory, forget it. You have no way of knowing who's a qualified buyer and who's not.

5. *Make it easy to respond.* Always ask for the order in such a way that it's easy for the buyer to respond. Make the order form big enough to write on. Enclose a return business-reply envelope with postage paid. Offer credit terms (free trial offer, internal charge card, no interest, COD), charge-card privileges, and a money-back guarantee. These will ensure a good response.

6. *Tell your story over and over.* Tell them the benefits of buying your invention in many different ways within the body copy. Reinforce the smart choice they're making by buying your invention. People need to be sold on the fact that they made a wise decision.

7. *Research every mailing you do.* Once you have finished a mailing, go over the results, and analyze what you did right and wrong. Compare with other mailings of similar companies, by checking textbook examples or calling your industry association. Read books on direct-mail and mail-order marketing, and get other ideas.

The Final Key.   If I had to pick one final key, it would be: Test, and test, and test again!

For additional sources of information, see William Cohen's *Building a Mail Order Business*. This encyclopedic volume contains almost everything you'll ever need to know about mail-order and direct-mail marketing.

NEWSPAPERS

Newspapers are the major advertising vehicle for most small businesses that market to restricted geographic areas. Since you want your invention to be exposed on a nationwide or even worldwide basis, however, you will probably find this medium to be of secondary advertising importance. (Don't forget, though, that newspapers are a primary means of getting free publicity.)

Just in case you do intend to do some newspaper advertising, you should know the basic way newspapers work and how to go about advertising in them.

Newspapers can be very good low-cost tests, *if* you have a consumer invention that the majority of the readers need. In order to reach these people, you must advertise in the publications they read. Here's where your target market research comes in. Find out who your customers are and what they read, research these publications, and pick the dominant one for a test.

The Pros

There are major advantages to advertising in a newspaper. While we normally talk about local papers, most of the advantages cover regional and national papers as well.

Quick Response Time. In today's world of advertising, you can't afford to wait a long time for the results of your test. The idea in target marketing is direct-response advertising, where you know *fast* whether your potential customer is interested in your invention or not.

The old image type of advertising campaigns that take forever are not for you; you have neither the time nor the money to wait. Most newspaper ads can tell you within a week to 10 days if your invention has a chance.

Short Lead Time. Most newspapers have a very short lead time. Lead time is the time period between turning in your ad and publication of the newspaper. Lead time on dailies can be as short as a few hours; lead times on weeklies, 1 or 2 days.

Ability to Merchandise Your Ads. You can price your products within the ads, change the product mix around, and juggle price according to reader response. This is impossible to do with yellow pages and major magazines, because lead time for the former is 1 year and for the latter, 2–3 months.

Habit-Forming Potential. Once readers are hooked on a publication, they make it part of their routine. It becomes a habit. If your ad is in the newspaper often enough, it might become a habit for regular readers to see what products and/or prices you are featuring today. Remember, it takes 21 days to make a new habit, so don't expect readers to buy from your ad immediately. Change and decision making take time.

Ability to Move Merchandise. When people decide to go shopping, they normally look in the newspaper first for store and product sales. With your ad in the paper, chances are very high you will attract a percentage of these shoppers.

Low-Cost Tests. As an inventor, you will be most concerned with advertising your invention on a nationwide or worldwide basis. But consider starting out with a small local test ad in a neighborhood paper for a few weeks. This should give you some insight into how well your invention will be initially received. (Do this *after* you have saturated the market with your PR campaign.)

Shelf-Life. People will refer to your ad. The longer the shelf-life of a print media vehicle, the more readers will reread your ad. (Shelf-life is the length of time people keep a publication around.) If they don't see the ad the first time, they will the second or third.

Flexibility in Placement and Timing. With newspapers, you have the option to place your ad anywhere you choose, within reason. You also decide when your ad should run. If you want the business section on Tuesday, just tell them. Unless the paper is a large metro edition like the *L.A. Times*, they probably won't charge you for special placement.

Co-op Advertising. In co-op advertising, the manufacturer pays for a portion of the ad and the dealer pays the rest. If you are a retailer and can get enough of your manufacturers to joint-venture an ad with you, your cost should prove quite nominal. I recommend you pay the least amount possible, and let the others carry the burden. (As an inventor, you may be asked to foot the manufacturing bill.)

The Cons

The cons of newspaper advertising are far fewer than the pros.

Clutter Problem. Newspapers have the same clutter problem as other media. There seems to be too much advertising these days, and not enough space devoted to editorial. The normal ratio used to be 60% advertising to 40% editorial. That seemed like a lot. Recently, many newspapers have changed their ratio to 70:30—70% advertising, 30% editorial! And the trend seems to be getting worse.

This can put *your* ad in a bad atmosphere, as it were. If the paper is too cluttered, people might bypass your ad or the whole paper itself. A perfect example is the Saturday sports section of most major dailies. Hundreds of computer companies place their ads in this section every week. This "follow the crowd" mentality is detrimental to their image. Such advertising implies cheap, discounted products.

Skimming. If readers are in a hurry, the editorial may attract them, not the ad. People skim a publication by the headlines. If the editorial headlines say "Read me" but your ad's headline doesn't grab their attention, they'll skip your ad entirely.

Poor Reproduction Quality. The reproduction quality of ads in a newspaper is poorer than that in a magazine. Black-and-white photographs reproduce very well, but color does not. If your invention *must* be shown in color, don't advertise in a newspaper. Some ads do not reproduce well at all. Check with your local newspaper's art department for help.

Which Newspapers to Use

By now it should be obvious I recommend local papers for an *initial test*. They're inexpensive and results can be measured very quickly. But which locals should you start with? There are several different types: suburban, city, and special-interest. Also consider the daily and weekly. (I don't recommend Sunday papers or shoppers—too much clutter.)

Earlier in the chapter, we examined the characteristics of the main types. I advise you to start with suburban dailies or weeklies, then move up to city weeklies, then city dailies.

Special-interest papers, if within the realm of your industry, should also be strongly considered. If there are such entities (more like trade magazines than newspapers), start with them first. If not, follow the other order.

Circulation vs. Readership. One of the considerations in making your decision about which newspaper to use (after you've determined dailies or weeklies) is circulation. Circulation is a very funny game. Newspapers and magazines play with the figures all the time. It's up to you to understand what sales reps are saying, and pin them down on the actual figures.

For example, a paper might say they have a circulation of 20,000. That may seem small, but consider several factors. Is the paper a weekly or suburban? Most have circulation within this range. Is the town the paper serves small, either geographically or demographically? Depending on various factors, 20,000 could be either large or small.

What is meant by circulation? Is it the number they print, or the number they print and put on the street? It should be the latter. And what about all those papers in the racks and on newsstands? How many are returned to the plant unsold? You need to know all of this. If a paper circulates 20,000 copies but only 5,000 get sold, you're paying for 15,000 wasted copies.

Does the sales rep talk circulation or readership? There's a *huge* difference. Circulation is the number of papers distributed. Readership is the number of people who *read* it. *They're not the same!* Two or three people may read the same copy of a paper. This is called the pass-along factor. If the sales rep quotes readership numbers, don't automatically be impressed. The rule of thumb is that there are 3.4 readers for every paper printed. Divide the sales rep's number by 3.4 to get the true circulation figure.

How Often to Advertise

Someone once said you should advertise "often enough to get the job done!" This doesn't tell you anything. According to Jay Levinson, author of *Guerrilla Marketing* (Boston: Houghton Mifflin, 1989), it takes nine impressions for an ad message to sink into your prospect's brain. Since a reader sees only about one of every three ads you run, theoretically you need to run at least 27 ads to get "normal" response. Levinson says you must have patience in order for advertising to work. I agree.

Types of Newspaper Ads

There are four types of advertising in a newspaper: general display, retail display, classified display, and classified ads.

Display advertising is something you are probably very familiar with. Any time you see a department store ad or a product ad in a newspaper, you are looking at display advertising. They are advertisements using attention-getting elements, such as artwork, illustrations, various type styles, sometimes color, or white space, in contrast to classified ads, which are made up of one style of copy in small, plain type.

General Display. General display refers to the rate card (price list of ads) involved in the purchase. General display is national or nonlocal advertising. The rates are usually much higher than retail rates; at least 20% more in some cases. This is supposedly to make room for advertising-agency commissions, which come into play during national ad buys. (I could never understand why the general rate card at a former employer of mine was 27% higher than the retail rate cards. Using a standard 15% agency discount, this gives the paper 12% more profit.)

Media will try to get the most they can for their advertising, and they may try to pawn off a general (national) rate card on you. There are two things you can do to prevent this:

1. Tell them you want a retail (local) rate card.
2. Tell them you buy your own advertising. If no agency is involved, there's no reason for a general (national) card.

Retail Display. Retail display advertising is exactly the same as general display in terms of space. The difference, as noted above, is about a 27% lower cost to you. Always ask for a retail (local) card when dealing with a space rep.

One more thought on general vs. retail rate cards. I used to work for a regional computer publication. I once asked our publisher why manufacturers and agencies pay on a higher rate card, since it's all the same space. A quarter-page is a quarter-page is a quarter-page, right? Looking at me like I was from Mars, he said, "Manufacturers have more money to spend on advertising; they expect to pay more on a higher card." I must be dumb. I still don't understand the difference—do you? (Watch yourself!)

Classified Display. If you look in most general newspaper classified sections, you'll find small space ads within the classified columns. These can be either agate (classified) type or regular display type. They have a thin border around them and fit within exact column dimensions. These are classified display ads. They cost less than general or retail displays and a little more than classified ads. Normally you are charged by the column-inch (1 column wide by 1 inch deep).

Classified Ads. These are usually found in the back section of the newspaper. They are set in agate type and are listed by category or classification. They can be effective if the paper has a large enough circulation. Therefore, I only recommend you use them in large metro dailies or national publications.

If you want to do classified advertising and make a strong impact, try using classified displays. For a little more investment, you stand out from the crowd. And really, that's the name of the game.

Rate Card Fiction

Besides general and retail rates, be aware of the other ways that newspapers price their space. There are flat rates and discount rates.

Flat Rates. A uniform rate for advertising space or time, with no discounts for volume or frequency, is a flat rate. Volume refers to the size of the ads. Frequency is the number of insertions you intend to run.

Discount Rates. Newspapers and magazines normally discount the cost of ads, based on two factors. If you run a larger ad, your *cost per column-inch* is less. If you run more insertions within a given time period, the *cost per ad* goes down. Here's an example showing how this works.

A newspaper has a total of 1000 column-inches on a page. The half-page cost is approximately $4000. The full-page cost is $7000. Column-inch cost is the full-page cost divided by the number of column inches:

$7000 ÷ 1000 = $7 per inch

A half-page ad costing $4000 has a column-inch cost of $4 per inch. Since two half-pages equal a full page, you would think the cost for a full page would be $8000. But it's not. As an incentive to buy a bigger ad, the newspaper discounts the larger size by $1000, or $1 per inch.

The cost per ad, or *frequency discount,* is a reduction in the ad cost based on the number of insertions used in a given period. If you agree to run ten ads per week rather than two or three, your cost per ad is less.

Ad Positions

There are two categories of ad positions within a newspaper: ROP and preferred.

ROP. ROP means "run of paper." In this case, the position of an ad is at the publisher's discretion. You pay your normal cost for the ad, with no premium charges added on.

Preferred. Any ad position in a publication where you have to pay a premium when you order it is called preferred. Some papers charge 10%, some 20% more. (This charge is a percentage of the space cost per ad, not the entire agreement.)

You can order a premium position ad by designating either the page or the location on the page. If you want an ad on page 15, that's the first way. Telling a paper to put you on any page, but in the upper right-hand corner, is the second way. The call-page method is usually the most expensive add-on charge.

Sections. Since all papers have separate sections for business, sports, women's news, entertainment, and others, it's best to pick the one for your ad most likely to draw the greatest response. This may be hard to do if you have multiple audiences. If you know your number-one target audience, and you should, try the section those people tend to read first.

Negotiating Position. If you are a frequent advertiser, pay your bills on time, and don't hassle the rep too much, you are an ideal account for that paper. Now's the time to leverage that relationship. Tell them you want to be up front in the paper, right-hand page ad, no premium cost. If they say it can't be done, try persuasion. "Oh, come on. I'm a good advertiser. What can you do for me?" This approach works often enough to try, but don't do it regularly. Be creative.

Color

Newspaper color has gotten more sophisticated as the years go by. But they're still a long way from the quality of magazines, and never will be as good. Printing on newsprint is different than printing on slick coated stock.

If you intend to use color in your ads, make sure it's not cartoon color (the colors used in the Sunday comics). Use the color to highlight the ad, not to overwhelm it. Newspaper printing colors are process cyan (blue), process magenta (pink), and process yellow. Stay away from yellow, unless you see a printed sample first. Sometimes newspaper yellow looks pretty bad.

You can also have premixed colors (called PMS colors) in your ad, but they tend to be more expensive than process colors. Check with your newspaper rep.

MAGAZINES

Magazines will be one of the most important aspects of your advertising campaign. You will probably advertise more in magazines than in newspapers because as a rule, inventions need to be marketed nationwide.

Categories

Magazines can be divided into three categories: trade, business, and consumer.

Trade Magazines. Trade magazines are for specific industries. They furnish news, reviews of new products, application articles, product roundups, and other industry-specific editorial information. A good example would be the computer industry. If your invention involved computers, you would have *over 400* national trade magazines alone to choose from!

Business Magazines. Business magazines are designed for the business community within certain demographic or geographic boundaries. For example, *Inc. Magazine* caters to small businesses, while *Entrepreneur* and

Success are for entrepreneurs. These are demographic classifications. Examples of geographic business magazines are *California Business* and *San Diego Business*.

Consumer Magazines. This is the broadest category; there are over 16,000 in the United States alone. They run the gamut from women's magazines to sports, from religious to romance, and from contemporary to clubs or organizations.

As an inventor, your first choice for publicity would be trade magazines, with consumer and business magazines to follow. In terms of advertising, depending on your invention, you would most likely choose an appropriate consumer magazine for mail-order or direct-response advertising.

Check with *Standard Rate and Data Service's* Trade Magazine, Business Magazine, and Consumer Magazine volumes to fill in your advertising contact cards.

Regional and local magazines can be beneficial if you have an invention that appeals to their audience. These would include magazines like *Los Angeles, New York, Boston*, and so on. Classified throwaways are those you find jammed into your mailbox every day. They are like the *Pennysaver*, with discount merchandise—not the vehicle for your product or service.

Characteristics of Magazines

The chief advertising characteristics of magazines are market selectivity, key customer coverage, long shelf-life, and quality look and prestige.

Market Selectivity. Magazines are perfect for market selectivity, another term for target marketing, the kind you will be doing throughout your campaign. Start with trade magazines. In order to get the word out to mass buyers and wholesalers, pick the key trade journals from the *SRDS* volume on Trade Magazines. What could be easier? Make a list of all the possible target markets by industry, look them up in *SRDS*, rank them according to circulation, and send for media kits.

Consumer magazines are also determined by market selectivity. In this case, use the *SRDS* volume on Consumer Magazines, as well as *Writer's Market*. Here's an example.

Let's say you've invented a new antitheft purse attachment for women, one that absolutely guarantees the purse cannot be snatched. Your first question is, "Who would be interested in this product?" Obviously, women. But women can be divided into many categories: teenagers, seniors, businesswomen, housewives, etc. In the *Writer's Market* category General—women's, you see the following magazine classifications:

Bridal magazines
Country women's magazines
Black women's magazines
High-fashion magazines
Working women's magazines
Homemakers' magazines

This may seem like breaking it down into ridiculously small subdivisions, but that's what you have to do in order to get a handle on targeting properly.

You start with working women's magazines, since you've decided most working women have the greatest risk of purse snatching. Upon examination, you find that the following magazines might be good vehicles for your ads:

Woman Magazine—Circ. 600,000
Working Mother Magazine—Circ. 600,000
Working Woman Magazine—Circ. 900,000

Because *Working Woman* has the largest circulation, you start with an analysis of it for possible advertising. And of course, you send for media kits from all three.

Key Customer Coverage. This ties in with market selectivity (target marketing). You must be sure to reach your key buyers. It's not enough to target via a subclassification, then pick the circulation. One of the points for analysis in the above example would be: Who are these working women, and do they have the income to purchase my invention?

You'd find this out by studying the readership survey the magazine sends in the media kit. It should tell you all the demographic data you need to make a reasonable decision: age, income, type of jobs, income level, etc. Combining this information with circulation and one more factor, you'll be able to pick the magazine to advertise in.

Long Shelf-Life. As you know, shelf-life is the amount of time a publication is kept in a business or household. The longer the shelf-life, the more cost-effective the ad. If a magazine is weekly, the shelf-life of 1 week means you have to pay four times the cost of an ad to get 1 month's worth of coverage. On the other hand, a monthly magazine with a 3- or 4-week shelf-life gives you more for your money.

Quality Look and Prestige. Magazines have an aura about them that says quality, prestige, class. When you advertise in a magazine, that type of image rubs off on your company and invention. And you can use it to your distinct advantage. It gives you credibility.

Key Points in Magazine Advertising

1. *Hit your target audience.* I've already talked about this in several places. It is important to note that this is the first factor in the proper use of magazines in your advertising campaign.

2. *Advertise in regional editions of nationals.* Most national magazines have regional editions that you can advertise in at far lower cost than the entire national run. Yet people aren't aware of this, so they'll think your ad was seen nationwide. Look at the leverage this gives you. If you buy ads by region one at a time, you can test in a national magazine and save money in the process.

3. *Mention your magazine ads in other promotions.* Get reprints of your ad from the magazine, or make your own. Cut out the ad page, reduce it, then reproduce it on your letterhead with the words "As Seen In" displayed at the top. Send these reprints out in a direct-mail package, along with a publicity release, or any other type of promotion. This is marketing at its best—paying for an ad once, then having it do triple and quadruple duty.

4. *If you're on a small budget, do the two-step.* If your budget is tight, run a small space ad in a regional edition of the first magazine you chose.

In your ad copy, make sure you invite readers to send for more information. When they do, send a brochure and sales letter as follow-up. You can still do a "As Seen In" reprint on your stationery. Just enlarge the size, so it looks much bigger. Most people won't know the size of the original ad.

The Pros of Magazine Advertising

1. Great pass-along rate.
2. Long shelf-life.
3. Strong reread factor.
4. Vivid color and creative effects.

Pass-along rate and shelf life were previously discussed in the section on newspaper advertising. The reread factor is the number of times a potential customer will go back to reread the magazine and possibly look at your ad.

The Cons of Magazine Advertising

1. Long lead time.
2. Limited opportunity for a quick response.
3. Cold leads (longer than 2 weeks).
4. Audience duplication.
5. It is difficult to sell items directly to the reader, via mail order, since the invention is an unknown. Image ads, such as "Buy this type of product somewhere," work better. Don't expect a magazine ad to sell your invention directly from the ad.

Magazine Rates

Rate cards for magazine advertising should be watched as closely as those for newspapers. Remember that rates are not gospel; they are determined by human beings, often with no formula for how they determined a certain rate. Everything should be negotiable.

Circulation, discounts, and premium positions are just as important with magazines as with newspapers—maybe more so, since you'll count more heavily on using them for your advertising messages.

CPM (Cost Per Thousand)

This method applies to both newspapers and magazine rates. In order to determine which publication is giving you the lowest cost, agencies came up with the CPM, or cost-per-thousand method. As a simple example, how much will it cost to reach 1000 people? The lower the cost per thousand (assuming two publications have the same circulation, or rate base), the better a buy that particular publication is.

Simply divide the full-page open (1X) rate by the number of thousands of circulation. For example, a magazine with 100,000 circulation and a full-page open rate of $2395 has a CPM of $23.95. In other words, it costs $23.95 to reach every 1,000 readers. If another magazine has the same circulation but a higher full-page open rate—say, $4000—the first magazine is a better buy.

Magazine Networks

Many magazines are part of so-called networks, or loose affiliations of different publications that try to help each other in selling advertising. Sometimes you can get a better deal by going through the network.

In southern California, for example, *Los Angeles Magazine*, *Orange Coast Magazine*, and *San Diego Magazine* have put together their own network. If you were to buy an ad in each one separately, it might cost you more than you could afford. However, if you were to go through the network and buy ads in all three at one time, you'd probably save at least 10%, if not more.

Before you do any advertising in multiple cities, check *SRDS* and *ADWEEK* to see if your choice's are part of a network.

BROADCAST MEDIA

The two components of broadcast media are radio and television.

Pros and Cons

Each medium has distinct pros and cons. The pros of radio are:

- It's an intimate medium.
- It's a personal medium; the listener is likely to hear at least a portion of the message.
- It's more targeted than television, less targeted than direct mail.
- It has more flexibility in terms of time and money.
- It offers opportunities to quickly plan special events.
- The repetition of your message reinforces your image.

The cons of radio are:

- Lots of the time radio is background noise.
- People can't remember the commercials very well.
- It's hard to pinpoint the radio market geographically.
- You can waste money because of audience overlap.
- You rely too much on station personnel to create your ad.

The pros of television are:

- It has visual impact.
- Its action orientation.
- It's an effective means of demonstrating products.
- It can be glamorous and exciting.
- It makes you look bigger than you really are.
- It creates uniqueness and reinforces memory.
- It creates credibility in the viewer's mind.

The cons of television are:

- It can be very expensive.
- It can become complex.
- The viewer cannot review the message (no reread factor).
- People are watching less TV today.
- Viewers may become preoccupied while watching.

Key Questions About Broadcast Advertising

Before you venture into the realm of radio and television advertising, you must answer some key questions.

1. *Should I select an AM or FM station?* It used to be that AM stations played rock and roll music, had news and sports, and were the low end of the advertising spectrum. FM was considered more upscale. But with today's radio stations playing all sorts of formats on both AM and FM, the distinction is blurring. Don't choose a station based on what band it's on. Rather, choose based on whether or not it reaches the correct audience.

2. *VHF or UHF television?* UHF television has come a long way since the 1960s, when it first came out. However, there is still a class distinction. VHF television includes the major networks and independents, while UHF includes high-band independents. Depending on your product, you might consider UHF (mail-order products, mostly). Otherwise, stick with the main channels on VHF.

3. *Who is listening and who is watching?* This is the most important question of all three. Again, we go back to our earlier discussions of targeting the audience. Does the station reach those I want to reach? If yes, I will *consider* them for possible advertising. If not, no salesperson will change my mind!

Radio/TV Ad Classifications

As with newspapers, the broadcast media have their different rate cards. You can buy a local flight (a campaign running a specified number of weeks), or a spot buy. Local buys are commercials only in the local market. Network flights are nationwide buys on all the affiliates of the network. Spot buys are commercials purchased on a market-by-market basis.

Cable vs. Regular TV

Cable television has come a long way from the days when it was just used for improved reception. Entire cable networks are now in operation 24 hours per day (thanks to visionary Ted Turner).

For kids, there's Discovery Channel or the Disney Channel; for teenagers—give them their MTV; for sports nuts, there's ESPN, Prime Ticket, and the Sports Channel; for women and homemakers, there's Lifestyle. No matter who's your target market, you'll find a cable channel that's right for them.

Determine who you want to reach, call your local cable ad rep, and get their media kit. You'll be surprised at the prices.

I recently produced a 30-second spot commercial for an auto engine rebuilder for the total price of $900, including scripting, raw footage, music, and finished product. (We got to keep the raw footage, which will be used in future spots.) The cost per 30-second spot was just $25, and we're targeting directly to the market he wants—what he calls the "weekend warrior" do-it-yourselfer. For $1300, the cost of one 30-second spot on the local ABC affiliate, we bought ESPN and Prime Ticket for 8 weeks!

Our negotiation technique was pretty good, too. The rep was sure Discovery Channel would help; I wasn't. So I said, "Prove it!" They gave us, at last count, 260 free spots on Discovery to prove it was a market for us. (This technique is called matching avails, or availables. Ask for them *after* your rep has offered the lowest rate possible, and you've said OK.)

Spot Announcements

There are various types of commercial advertising buys in the world of radio and television, in addition to the 30- and 60-second spot commercials. How you buy them—that is, how much you pay and how they are classified—is very important. Here are some definitions you should know about buying spots.

Fixed Spot or Rate. This is the maximum rate paid by an advertiser for a spot to ensure that it runs in that position without being bumped by a higher paying advertiser.

Preempted Spot. This is a rate or spot subject to cancellation by another advertiser who pays a higher rate. There are varieties of preemptive rates: On 2-weeks notice, on 1-week notice, or no notice. Each station varies in its protection period. Be sure you check the contract and understand all the rules if you buy this type of commercial.

Floating Spot. For a floating spot, you pay a little more than for a preemptive spot and less than for a fixed spot. It is similar to a newspaper ROP.

You know you'll have a spot somewhere on the schedule, you just don't know what time; it varies from day to day.

ROS Spot. ROS stands for run of schedule. This is a commercial spot that can be scheduled at the station's discretion anytime during a period you request. In other words, you specify the period between 10:00 A.M. and 4:00 P.M. The time the spot runs will vary daily.

Package Plan. In radio, this is usually the purchase of a sufficient quantity of spots to entitle you to a discount.

Commercial Length

Commercial time segments are getting smaller and smaller. Television pioneered the 15-second commercial a few years back. Now there is talk on Madison Avenue of a series of 5-second commercials. Can you imagine that? That's twelve spots in 1 minute! Viewers won't remember anything. Steer clear of these!

In my opinion, with some exceptions, like certain television commercials for products that need major demonstrations, no commercial should be longer than 30 seconds. If you can't say it in 30 seconds, you're not saying it right. (This advice will save you money, too—a 30 is 70% the cost of a 60.)

In scripting, you can say twenty words in 10 seconds. So a 30-second spot will take up sixty words. You can write a nice little spot in sixty words, believe me. Just mention the top two selling points, why they are benefits to the viewer or listener, mention your invention's name several times, and give the phone number. That's all there is to it.

Keys to Radio Success

1. Produce an attention-getting opener.
2. Introduce the invention's benefits.
3. Use a "me to you" tone of voice.
4. Use clear and crisp language.
5. Make sure you have strong name identification.

6. Include testimonials.
7. Use an action-getting closer.
8. Focus on one invention per spot.
9. Add music for instant identification.
10. Make sure the words attract the listeners.
11. Sentences must be simple.
12. Repeat your invention name often.
13. Read your spot aloud for clarity.
14. Your spot is a sales pitch—remember that.

Tips for Good TV Commercials

The keys to radio success also apply to TV. Television production costs can be kept down if you remember these tips:

1. You don't have to compete with the "big boys."
2. TV is *not* show business.
3. Studios can use inexpensive color slides for the visuals.
4. Use your local cable company to produce your spots.
5. Commercials can be used over again and the costs can be amortized.

Time Segments

Both radio and television have standardized the daily time segments. Each medium does it slightly differently. Both set rate cards by the popularity of broadcast listenership and viewership.

Radio Time Segements

Morning drive time	6:00–10:00 A.M. (Expensive)
Mid-morning	10:00 A.M.–Noon (Moderate cost)
Mid-day	Noon–3:00 P.M. (Medium expensive)
Afternoon drive time	3:00–7:00 P.M. (Expensive)
Evening	7:00 P.M.—Midnight (Moderate cost)
Graveyard	Midnight–6:00 A.M. (Cheapest)

Television Time Segments

Morning	6:00–9:00 A.M. (Moderate)
Mid-day	9:00 A.M.–4:00 P.M. (Inexpensive)
Early fringe	4:00–7:30 P.M. (Moderate)
Prime time	7:30–11:00 P.M. (Most expensive)
Late fringe	11:00 P.M.–1:00 A.M. (Moderate)
Graveyard	1:00 A.M.–Sign-off (Cheapest)

These time sequences can vary up to 1 hour.

How to Buy Spots

The name of the game in buying commercials is negotiation. Here are a few tactics.

1. For radio, early in the week is a better buy, because the number of available spots is higher. Most companies advertise at the end of the week, when the buyers and payday comes. If you have an invention or product that can be advertised early in the week, you'll save money.

2. Tell your rep your budget up front, *then* negotiate for the number of spots. He might throw in a few extra for you, since mentally he's pocketed his commission.

3. Dominate a single station, rather than spreading your spots out over several different ones. Use your entire weekly budget, and buy the best station in terms of CPM. You'll look like a heavy hitter.

4. Make a friend of the traffic coordinator. This is the person who logs the spots onto the computer. They are overworked and underpaid. Always thank them for their help. Once you establish a relationship, you'll probably get better time periods.

Finding the Right Media

The techniques for finding the right media are very similar to those described in Chapter 12 on publicity. You will use basically the same sources: media directories, association directories, and the media themselves.

The first directory to use would be *Standard Rate and Data Service's* various volumes. One is on consumer magazines, another on direct mail, still others on radio, television, and newspapers. It is *the* source for ad rates. Look in the index for your particular industry category. Make a list of all possible media that might be appropriate places to advertise. (You should have already done this in your prepublicity research.)

Next, write to the media themselves and ask for media kits. This is the advertising version of their press kit. It will include everything they deem necessary to "sell" you on using them.

At this point, don't be intimidated with the amount of material you begin to receive. You will sort through the volumes of information quite easily, and note this information on a second set of media contact cards. (It will be necessary to have a second set of cards, in addition to the ones you developed for publicity. However, there will be fewer cards because you'll only include specifically targeted media for advertising.)

When you receive the media kit, look first at the readership or broadcast survey, which shows what the audience is like. If it doesn't match your target market, put the information packet aside for now. If there is a match, however small, put the packet in a pile marked "Possible Advertising Vehicles." Next, look at the rate card. It tells you the price for various ads or commercials. In print media, it will show either cost per ad (magazines) or cost per line or column-inch (newspapers). In broadcast media, the rate card will show cost per spot, package rates, or sponsorships (see below.)

Finally, outline the information you receive on a media cost-comparison grid (Figure 13.3). You can see at a glance what each method will cost you, and where you might spend your money.

Typical Magazine Rate Card

	1X	3X	6X	12X
Full page	$1200	$1150	$1100	$1050
½ page	900	950	900	850
¼ page	750	700	650	600
⅛ page	450	400	350	300

This fictitious rate card for magazines shows rates at one insertion, three insertions, six insertions, and twelve insertions. The rate for a full-page, three-time run would read $1150 *per insertion*. Frequency discounts show a $50 difference. The norm is a 5–10% discount per insertion category.

Print Media

Medium	Circulation	Cost Per Thousand (CPM)	Cost Per Ad
1.			
2.			
3.			
4.			
5.			
6.			

Broadcast Media

Medium	Audience Per Quarter Hour	Cost Per Spot	Package Cost	Sponsorship Cost
1.				
2.				
3.				
4.				
5.				
6.				

FIGURE 13.3 Media Cost-Comparison Grid

Typical Newspaper Rate Card

	Daily	*Sunday*
Open Rate	$65.00	$69.00
50"	52.26	55.67
150"	39.13	42.46
250"	35.27	38.43
500"	34.94	38.09
800"	34.60	37.75
1200"	33.93	37.42

This fictional rate card for newspapers shows cost per column-inch. A column-inch is 1 column wide × 1 inch deep. To determine the cost per ad, multiply column-inches × depth × rate per inch. Rates are determined by the number of inches contracted for (first column in the above sample). For example, an ad 2 × 4 means 2 columns by 4 inches equals an 8-inch ad. The total cost on Sunday, if you have a yearly contract, or 1200 inches, is $299.36 (8 × $37.42).

Typical Radio Rate Card

	1	2	3	4
30 Second Spot	$100	$125	$75	$150
60 Second Spot	175	200	150	225

The numbers in the top row, called "grids," correspond to quarters of the year (1 = winter, 2 = spring, 3 = summer, 4 = fall). The lowest rate falls in grid 3 (summer), when most people are on vacation. The highest rate occurs in grid 4 (fall), just before the Christmas season. As you can see, rates for radio time fluctuate with the seasons and the buying mentality of the country. A wise advertiser will order radio spots during the winter quarter well in advance of the fall season, or lock in low rates by ordering an entire year's worth of spots.

Sponsorships

Includes two 30-second commercials and two 15-second commercials per hour.

1	2	3	4
$225	$275	$175	$350

Again, the numbers indicate the grid, or season when the spot is to run.

SUCCESSFUL AD LAYOUTS

In order for an ad layout to be successful, you must have the following:

1. Dominant element—usually artwork or a headline.
2. Alignment of elements—everything carefully aligned.

3. White space—the more white space, the better.
4. Placement of illustrations—face into body copy.
5. Typefaces—not too many in one ad.
6. Harmony of elements—all in balance.

MAIL-ORDER ADVERTISING

Mail-order advertising is very different from direct mail, yet many people get the two confused. Direct mail is outgoing; you send a solicitation via the mail. Mail order is incoming; you place an ad in a publication or on television and wait for the response. By following the ten rules below, you should be successful with your mail-order advertising.

Rule 1: Know the Mail-Order Rule

The Mail-Order Rule was issued by the FTC (Federal Trade Commission) to correct problems with late or undelivered mail-order merchandise. Any advertiser worthy of his reputation will follow it.

> Under this rule, you have a duty to ship product on time. You must also follow procedures if you cannot ship on time. When there is a delay, you must notify your customers of the delay, and provide an option of agreeing to the delay, or allow them to cancel the order and receive a prompt refund. For each additional delay, your customers must be notified that they must send you a signed consent to a further delay or a refund will be given.*

Rule 2: Know What Products are Right for Mail Order

Almost any type of product can be appropriate for mail-order marketing. Chinook salmon from Alaska, peaches from Oregon, cheese from Wisconsin, even luxury cars and boats have been sold through mail order.

Source: *Business Guide to FTC Mail Order Rule*, U.S. Federal Trade Commission, Washington, DC.

Understand, though, that the bigger the product and the higher the cost, the more ads it takes to sell via mail order. For most of your products, they can probably be sold via mail order. Just remember to make sure the items are familiar to the public; they should have a strong appeal, be practical, lightweight, and moderately priced.

Rule 3: Know How Much You Need to Spend

To do mail order properly, you must test your offer several times. This will involve an investment from several hundred to several thousand dollars. A good test would be at least three ads over a 3-month period (assuming the publication is monthly). If you spent only $20 on an ad and it returned nothing, even though the cost was minimal, it was a bad ad. On the other hand, if you spend $1,000 and it returns $10,000 in sales, the higher investment was worth it.

A general rule of thumb (although there really is *no such thing* anymore) is to get a return of 10 to 1 for every dollar spent. This takes time, however. Remember, you'll need to test different ads to see which one gets the best response.

Rule 4: Understand the Purpose of Mail Order

While most mail-order novices think the purpose of mail-order advertising is to sell the product they've advertised, professionals understand that the real purpose is to develop a mailing list for future appeal and rental.

This is why many people counsel businesses *not* to do any mail order at all. Some suggest that you contact the established mail-order houses and have *them* sell your product for you. But it's not as easy as it sounds.

So what do you do? Should you do mail order? Yes, provided, again, you have a product that is priced right. You might consider mail-order advertising to be a loss-leader item; you know initially you may lose some money in the promotion, but since you're going for the follow-up sales and list rental, in the long run you'll win out.

If you want a really smart way to do mail order, follow my suggestions for small budget magazine advertising. Test a few ads using the "Send for a free sample" approach. This way, you get names on your list, and those that respond have prequalified themselves one step further—they're not buying the product, but are asking for additional information. They know they're going to get a sales pitch back in the mail.

You will get a greater response initially, because there's no cost to them. Again, add them to your mailing list for a direct-mail campaign.

Rule 5: Knowing When to Do Mail-Order Advertising

After you have exhausted the majority of your efforts in a public relations campaign, begin to selectively test with a few mail-order ads. It would be wise to test those magazines that pulled the best response from publicity releases.

Rule 6: Use the Mail-Order Survival Guide

The mail-order survival guide can be used to determine if your product is right for mail-order marketing. The questions to ask are:

1. Is the product a repeat item or a one-shot deal?
2. Does it fill a real need? (Fads get expensive.)
3. Is it priced right? (Under $20 increases the risk of failure.)
4. Can it be packaged and shipped easily?
5. Can I make a profit after all costs are factored in?

Rule 7: Use the Correct Publications for a Test

Use targeted trade and consumer magazines with the largest circulation and lowest CPM. (See the section on magazine advertising earlier in the chapter.)

Rule 8: Use the Proper Type of Ad

While a classified in a national publication will pull response, a display ad usually does much better. The greater the visibility of an ad, the greater the pulling power. Display ads do just that—display your merchandise.

If you are using mail order as a lead-generation vehicle, you can start out with a classified. If you are using it for selling your invention, you *must* have a display ad with illustrations or photos (black-and-white) of your product.

Rule 9: Know the Proper Way to Write a Mail-Order Ad

In addition to everything you learned earlier in this chapter, there are a few concepts crucial to writing mail-order advertising.

1. Key the ad. Make sure you know where the response is coming from. You may do several different test ads at the same time. Use different phone numbers, post office boxes, department names (for example, "Write to: Inventor Magazine, Dept. IM389, 485 Madison Ave., New York, NY 10000"). This department code, Dept. IM389, translates to *Inventor Magazine*, March '89 issue. This is an excellent method of tracking your ads to see which pull the best response. Since the buyer writes to Dept. IM389, you know the magazine and the issue in which he saw the ad and decided to respond. If you run an ad in more than one magazine, coding is absolutely essential. Otherwise, you will become confused as to which magazine does the best job. For example, a two-magazine test would be Dept. IM389 for *Inventor Magazine,* March '89 issue, and Dept. PM389, for the same ad in *Popular Mechanics,* March '89 issue. Just count the number of responses sent to each "department," and you will know where best to spend your money.

2. Have a stopper headline. Mail-order ads *must have* strong benefit headlines.

3. Have selling body copy. Every word must sing a song of sales. Every phrase must be concise; no wasted words.

An excellent book on writing ad copy is Victor Schwab's *How to Write a Good Advertisement: A Short Course in Copywriting* (Los Angeles: Wilshire Book Company, 1980).

Rule 10: Make Sure Every Element Is There

A mail-order ad must contain every element that a direct-mail package would have: the offer, the return form, good layout and design, testimonials.

Take any direct-mail package you have run, and condense its elements into the shortest form possible. This will give you a starting point for your mail-order ad.

ADVERTISING RESOURCES

Here's a list of resource books I have on my library shelf. I suggest you add a couple if you're serious about advertising your invention properly. Of course, you could hire an agency or consultant to tell you what to do. But as you know, that requires spending more money.

> *Newspaper Advertising Handbook*
> by Don Watkins
> Wheaton, IL: Newspaper Book Service, Dynamo, 1983.

A great little book on the layout and design of ads.

> *The Advertising Answerbook*
> by Hal Betancourt
> New York: Prentice-Hall Press, 1982

Hal's a master of southern California advertising. A readable, understandable basic primer.

> *How to Advertise*
> by Kenneth Roman and Jane Maas
> New York: St. Martin's Press, 1976

One of the best books, with loads of check-off lists on what works best in print, radio, direct mail, television, outdoor, etc.

> *How to Make Your Advertising Make Money*
> by John Caples
> New York: Prentice-Hall Press, 1983

Caples was a master and taught more ad copywriters and agency people than anyone else. He includes sample after sample of what's good and what's not. A veritable encyclopedia of ad information.

14

Trade Show Secrets of the Pros

The trade show is the most misunderstood and maligned of all marketing tools you might use. If used properly, however, it will more than pay for itself. This chapter will explain in detail how to put on a successful trade show promotion.

TRADE SHOW GOALS

Trade shows have a number of benefits for the inventor-marketer. These include:

1. Low-cost testing *before* you spend lots of money.
2. Development of additional sales leads.
3. Attracting new dealers/distributors.
4. Development of new sales territories.
5. Recruitment of sales personnel and manufacturer's reps.
6. Introducing new products.
7. Identifying new markets.

8. Generating better public relations.
9. Educating buyers and sellers.
10. Solidifying relationships.

Whatever reason you have for exhibiting at or attending a trade show, make sure you thoroughly understand what that reason is and how you intend to achieve it. This is basic goal setting, which is the first step in proper trade show preparation. Success depends on two things: setting realistic goals and proper planning in order to meet those goals.

You cannot have more than two or three goals, or your marketing messages will get confused, and you won't be able to tell if you were successful or not.

Setting Objectives

Goals come first, then objectives. Consider that goals are the strategies of your trade show plan, and the objectives are the tactics. Write down your two or three goals. Then list under each one the steps for accomplishing that goal. For example, if your goal was to accomplish numbers 2, 3, and 5 above, you would write the following objectives:

1. Inform a minimum of 600 buyers of my new product.
2. Show the product in action to each one.
3. Give each one a piece of sales literature.
4. Get names, titles, and phone numbers for follow-up.
5. Contact 150 selected old customers, current hot prospects, and others who are desirable, and let top management [you] mingle with them.
6. Talk to and screen 20–25 manufacturer's reps.
7. See 25 distributors; invite them for a special screening.
8. Sign up 3 out of the 25.

Now that you have the objectives set, you can measure whether or not you are successful.

Planning the Show

Planning for a trade show takes thinking and hard work. After setting objectives, you need to determine the following.

The Show Schedule. Should you do national or regional shows? Which ones? Where do you get this information?

The Show Theme. What do you want to say? Who is your target market? How do they respond to different themes and stimuli?

The Show Design. Where should the exhibit be placed? What should it look like? How should it be constructed?

The Show Budget. How much is this going to cost? Is it in the budget? What are the elements of the budget?

THE MAIN INGREDIENTS

Determining the Show Schedule

Some books suggest doing only national shows, because of the number of buyers and lack of profitability at the regional level. But regional shows have become quite sophisticated in recent years and usually prove worthwhile for determining consumer and buyer response. Most regionals even take on a national scope, especially if held in major cities.

To find out which shows you might consider doing, get a copy of *Tradeshow Week Data Book,* available from *TradeShow Week Magazine,* Los Angeles, CA or your library. Each year's book lists all the scheduled shows alphabetically, geographically, by industrial classification, and chronologically. On your calendar, pencil in possible trade show dates.

The Show Theme

In Chapter 11 and elsewhere throughout this book, I have talked intensively about target marketing. Because trade shows are another marketing tool, the basic research we already discussed applies here as well. What is the profile of your prospective buyer? Write it down. Then ask yourself, "How do I translate my objectives for this group into concrete messages they will understand?"

Go back to your objectives list, and plan your attack around a core theme. If your tactics include sales leads, your theme might be "Whizbang Widgets Will Help You Sell More Profitably in the 1990s." (This assumes one of your key targets is manufacturer's reps.) For consumers, the theme might be "Save Money by Installing a Whizbang Widget."

You cannot have more than one theme simultaneously per audience and make it work. The central theme here is, of course, money!

The Show Design

Exhibit design is an art and one that I don't recommend you try. Find a good exhibit designer who understands marketing, not just art, and have a meeting. Explain your goals and objectives and what you hope to achieve over the year. Discuss theme, size of staff, traffic-flow patterns, booth budget, and other necessary items.

Putting these elements all together, have the designer prepare a rough sketch. Show it to your relatives, friends, and employees (if any). Take their suggestions, and rework the design. You should also visit several shows similar to the ones you intend to participate in, and get ideas of what others are doing.

The Show Budget

There is more to trade show budgeting than meets the eye. The logical way to develop a budget is to look at your objectives and determine what must be spent to achieve them. In our example, we have an objective of top management meeting with 150 cream-of-the-crop buyers. This will probably entail a cocktail party, dinner, or some other promotion.

Breaking this down further, we include cost of the meals, invitations, pre- and post-publicity, and any other items relating to banquets.

As you can see, from all the thoughts that go into a budget, this must be done months in advance of the show. Actually, you should budget for the year. Decide how many shows you would like to do. (At this point, this figure will simply be a wish-list.) Multiply each show by the number of objectives, what each objective will cost, and you get a rough estimate of a year's budget. If it's too excessive within your overall marketing budget, you'll have to cut back.

Understand the key factors that will throw your normal figures out of whack. If you have a new product, you will need a larger space and promotion money. If you need additional staff, their hotels, meals, and transportation will increase your costs. Freight, drayage (moving equipment to and from your booth), electricity, carpeting—all these add up. (For a list of budget items, see Figure 14.1.)

1. **Space Rental**
2. **Exhibit Fabrication**
    a. Design/redesign
    b. Construction/refurbishing
    c. Insurance
3. **Show Costs**
    a. Drayage
    b. Installation
    c. Rental of rugs, floor coverings, drapes, furniture
    d. Electrical
    e. Plumbing
    f. Flowers, plants
    g. Refrigeration
    h. Water, gas, compressed air
    i. Telephone
    j. Cleaning
    k. Photography
    l. Dismantling
    m. Shipping
4. **Arrangements**
    a. Rooms for personnel
    b. Suite entertainment
    c. Food/beverage
    d. Transportation
    e. Additional registration fees for personnel
    f. Guest passes
5. **Public Relations, Advertising/Promotion**
    a. Premiums
    b. Booth hoopla
    c. Direct mail
    d. Press campaign
    e. Photos
    f. Ads in trade magazines
    g. Ads in show program
    h. Pre-show gimmicks
    i. Miscellaneous
6. **Exhibit Staff**
    a. Company personnel
    b. Specialty personnel (models, demonstrators)
    c. Guards
    d. Transportation
7. **Company Material**
    a. Equipment for display
    b. Product for sampling or demonstrations
    c. Catalogues, brochures, sales literature
    d. Lead forms
    e. Installation and take down of equipment
    f. Shipping
    g. Blazers, uniforms for personnel
    h. Ad specialties
8. **Disposition**
    a. Reshipment
    b. Handling costs
    c. Exhibit revision
    d. Storage
9. **Other Costs**

FIGURE 14.1 Budget Items for Trade Shows

A rule of thumb: Assume you'll spend $5.50–$6 on the trade show for every dollar you spend on exhibit space. (This figure is an estimate. Prices vary according to region.) When estimating your trade show's budget, include the following:

1. Space rental: design, construction, insurance.
2. Show costs: freight, electrical, plumbing, carpets, phones, cleaning, shipping, photos, setup and tear-down.
3. Arrangement costs: hotel, food, transportation, fees, guest passes.
4. Promotion costs: publicity, direct mail, photos, ads, giveaways, literature.

THE EXHIBIT BOOTH

Your exhibit booth space is not the actual physical booth itself, but the floor space you are renting. In addition to design and execution, several other factors must be considered here. These are:

- Booth location
- Staffing needs
- Booth size
- Required elements for a successful booth

Booth Location

Where your booth is placed is one of the most critical factors in determining your success. Location can make or break you. The main considerations are proximity to the entrances, exits, and concession areas, and the traffic-flow patterns.

In selecting the right booth location, consider the following:

1. Booths near freight entrances require late setup and early tear-down.
2. If you use large quantities of paper or demonstration material, place your booth near storage areas.

3. Check floor plans in advance for obstructions that may prevent customers from reaching you easily.
4. Watch for changes in ceiling height. Some arenas don't allow high exhibit booths.
5. Understand the entire floor plan—where competitors are, aisle flow for attendees, etc.

Staffing Needs

Before you can determine how much space to buy, you must determine how many sales reps and technical support people you will need at the show on a per-hour basis. Although the formula is somewhat complicated, it works:

Total attendance estimated (TAE): _____

Paid attendance (PA): _____
    (Paid attendance = TAE − Media − Booth personnel)

Potential prospects (PP): _____
    (PP = 20% of PA)

Prospects attracted to my booth (PB): _____
    (Prospects attracted = 50% of PP)

Total show hours: _____
    (Total hours = # of days × 8 Hours/Day)

No. of visitors/hour: _____
(Visitors/hour = Prospects attracted/Total show hours)
    (Each rep can contact a *maximum* 15 people per hour with a 4-minute qualifying presentation, asking three quick sales questions to determine interest level.)

No. of personnel on duty per hour: _____
    (No. of duty personnel = Visitors per hour (VPH)/15)
    Multiply by 2 for rest periods

Total booth sales personnel needed: _____

To understand this formula, let's take the following example. Fill in the numbers as I go along.

Let's say the trade show promoters have told us the estimated attendance figures, based on last year's figures, will be 100,000. That seems like a pretty good number. We know we can't see all of them. But suppose we estimate we can see 10%, or 10,000 in a 4-day show. (I doubt that, but let's see.)

Fill in the TAE as 100,000. Now subtract the media personnel (0.5% is generous, or 500). The promoters say we're right on the money; last year, 500 media people registered. We deduct this figure from the TAE. We also deduct the number of booth personnel from all the other exhibitors. (Figuring conservatively 3 people per booth × 300 booths (number supplied by the promoter), we get 900 booth personnel + 500 media people = 1400 nonprospects.

Our PA, or paid attendance, is now 98,600. By industry studies, we should expect 20% of that to be potential prospects—that is, 20% of the paid attendance will probably be interested in our product category. That gives us 19,720 potential prospects (PP). Of this, if we go all out on promotion, we can at best hope for 50% of these to stop by our booth within a 4-day period. (Again, industry statistics. I feel this number should be lower, however.)

Our PB, or prospects attracted to our booth, is roughly 9860 in 4 days. Total show hours are 32 (4 days × 8 hours per day). Number of visitors per hour is PB divided by the total show hours. This comes to 308 prospects to our booth *every hour.*

If the number of personnel on duty is VPH/15, we need 20 sales reps on the floor every hour to do a 6-minute qualifying presentation. Multiply this number by 2 for rest periods, and you need 40 people just for sales.

If this seems unrealistic to you, it is. The purpose is to show you how far off the mark most people are. Either lower your number of prospects per hour, or increase the size of your staff. If you only have 3 people, reverse calculations show a *maximum* of 1440 prospects that can *effectively* be qualified as good leads in a 4-day show.

Booth Size

The size of your booth space is a very important factor. If you're too cramped, people will pass by. If you have too much space, you project an image of a bad product or company. The size-of-booth formula below will help you determine how much space you need. (Note: 100 square feet comfortably holds 4 people; therefore, 50 sq ft are needed for every 1 rep and 1 prospect. Remember that this is a selling situation in the booth; the more comfortable a prospect feels, the more likely they'll buy.)

Size-of-Booth Formula

Usable space: _____
(Usable space = # of sales reps on duty per hour × 50 sq ft)

Occupied space: _____
(Occupied space is all other space needed: exhibit, literature displays, sales stations, etc.)

Total space needed: _____
(Total space needed = Usable space + Occupied space)

Let's say you have five reps on duty per hour. Calculating at 50 sq ft per rep, you need 250 sq ft of usable space. You have another 300 sq ft of occupied space with the exhibit itself (furniture, sales stations, etc.). The total space needed is 550 sq ft. Since normal booth space is 10 × 10, or 100 sq ft, you need a booth with 600 sq ft (rounding off to the nearest 100), or a 20 × 30 booth. Calculate what you've been using; you'll see how cramped you have been.

Required Elements for a Successful Booth

There are many ingredients to a successful show booth. They include design, function, accessibility, and common sense. Here's the main checklist:

_____ 1. Strong brand-name visibility or identification signs.

_____ 2. One message per small booth; two or three messages per large booth.

_____ 3. Simplicity—don't detract with an overpowering structure.

_____ 4. Minimal graphics to highlight your theme.

218   *Part V   To Market, To Market*

_____   5. Subdued color for basic displays; contrast color for trim. Color-coordinate your table skirting with the carpet.

_____   6. Good lighting; you can never have too much.

_____   7. Display counters functioning as sales stations.

_____   8. An orderly, attractive appearance.

_____   9. Allow for good traffic-flow patterns.

_____ 10. Have a small storage area for tools, extra literature, etc.

_____ 11. Don't double-pad the carpet. It is actually harder on your feet and will cost you more.

_____ 12. Have enough seats for business use only.

_____ 13. Use motion in your exhibit through lighting, mechanics, staging, etc.

WHAT TO DO FIRST

_____   1. Read and reread the exhibit manual, especially the parts on hall regulations and labor union rules.

_____   2. Order all services and supplies in advance.

_____   3. Make a master list of all box contents, products, tools, etc.

_____   4. Remember: Display only the most interesting items.

_____   5. Make sure ID signs are visible at a distance of at least 100 feet.

_____   6. Have a main focal point for the booth.

WHAT TO DISPLAY

What you show at your booth is crucial to your success. It is impossible to display everything in your product line (unless you have just one item) on the chance that someone might want to see your new gizmo. Here are some suggestions for display.

New Products

This is one of the main reasons people exhibit at trade shows—to show off new products and get a feel for the marketplace.

Old Products

If you don't have a new product, bring out an old one, and give it a new twist. According to the Trade Show Bureau (the industry association), old products seem new to 80% of any audience; 60% of this year's audiences were not at last year's shows. Add to that 20% or more who did not get to see your booth, and you have the 80% who will love your "old saw."

You can also build your exhibit around a new twist, which can be exciting.

Sales Literature

While literature is fine for prospecting, and it does belong in the category of what to display, there is a proper way to use it.

1. Don't display literature for people to grab. These "literature-stuffers" and "looky-loos" cost exhibitors thousands of dollars each year. Store the sales literature in carousels or sales stands, and use it to qualify prospects. For those who just "have to have" some of your literature, prepare a one-page, inexpensive flyer giving the highlights of your products. But keep the expensive 4-color brochures *out of sight*.

2. Have more pieces than you think you'll need. A reasonable figure is 10% overage, just in case.

3. If more than one piece is being given to qualified prospects, prepackage them into a kit.

4. Remember to put your name, address, and phone number on *every* handout.

5. Plan your allocations according to the day. Most trade shows are not stagnant in terms of visitors. In a 4-day show, for example, day 1 will see about 20% of the total audience; day 2, 40%; day 3, 30%; and day 4, 10%. Be sure to check your quantities daily.

PUBLICITY FOR TRADE SHOWS

In Chapter 12, we discussed, among other things, the various methods for identifying the appropriate media, dealing with them, and preparing a news release. In trade show publicity, it's a *show*—it's *showtime!* A trade show is like a giant carnival. In addition to publicity, you should be thinking of different promotions you can do—before, during, and after the show.

Pre-Show Techniques

The more pre-show work you do, the better your show will turn out, and the less follow-up you'll have to do. Planning ahead works. Here are some ideas for pre-show promotions:

1. Prepare a *printed fact sheet* about your exhibit. Send it with a letter to your clients, prospects, and other interested parties.
2. Have an *in-house campaign,* where visitors to your operation can see what's going on.
3. Promote the show in your *daily mail* to all. Use a special *postage-meter imprint,* or add wording or stickers to the outside of packages.
4. Prepare a simple *show newsletter,* one sheet, printed on both sides, and send to all prospects and customers a month before the show.
5. Mail *show-supplied invitations* with letters to customers and prospects. Unfortunately, I've seen thousands of these left sitting in exhibitors' offices, unmailed. The promoter is wasting money, and the exhibitors are wasting a golden opportunity. Use them!
6. Make up a *show special* to give away at your booth.
7. Mail *discount attendance tickets* to people on your mailing list.
8. *Tag your ads* in newspapers, on radio, and on television prior to the show with the show name at the bottom. Sometimes the show promoter will co-op with you and rebate a percentage of the cost. You're helping him promote his show.
9. Send *news releases* about your products and participation in the show to both local and show-site media at least 3 weeks in advance.

10. Send *news releases* with unusual angles to members of the trade press in your industry.
11. Prepare *press kits* for the show.
12. Mail *reminder letters* to key accounts and prospects 1 week before the show.
13. Put a *map* of the hall on the back of all sales letters, and show the route from entrance to you. (You want the prospects to bypass the competition!)
14. *Advertise* in the same media the *same day* the promoter does.

At-Show Promotion

Your promotion effort doesn't stop once you get to the show. In fact, the pace should increase. You're there, it's exciting—time to let the world know about this excitement. Contact the local press in the city the show is in, and let them know you're in town and available for interviews. (You did contact them in advance that you were coming, right?)

Hold a daily drawing for a prize. Put a big cardboard clock on an easel, and announce the time the drawing will be held.

Meet with your local representatives and customers, and find out what help they need. Use the time at the show to find new reps and agents. You might even consider asking other exhibitors to represent your line.

Be sure to take plenty of photos. You will use these for post-show follow-up.

Post-Show Promotion

After the show, remind everyone what a great show it was, what a good turnout you had, and what a great job everyone did. Here are some suggestions:

1. Mail a follow-up letter to all who visited your booth.
2. Send a report to all your reps and dealers on what you did at the show and on how people responded.
3. Mail PR releases to trade magazines and local papers, telling them what a fine job you did and showing them with pictures.

4. Analyze what went right and what went wrong, and begin planning for next year.

LOGISTICS OF THE SHOW

Logistics involves budgeting (discussed earlier); organizing the personnel; keeping track of product, literature, booth, and supplemental items; and dealing with the various players within the trade show world. If you're a one-person operation, good luck! You've got to do it all. If you have several people in your company, appoint one person in charge of logistics.

The following are some of the concerns of the logistics manager.

Transportation

Transportation includes the transport of personnel, equipment, product, and anything else to the show site. Transportation is divided into two segments: shipping and drayage. Shipping is moving the stuff to the show site; drayage is moving the exhibit and equipment into the hall and to the booth location. The latter stage is usually handled by union workers.

Union Labor

Everyone has stories to tell about union problems at trade shows. It seems you can't get anything done without using union personnel. You pay good money for a half hour's time, they take 10 minutes, and then leave. Believe me, you can't fight city hall in this case. Your best bet is to cooperate with the union people. If there's a problem, refer it to the show's management. That's what you pay your booth fee for.

Installing and Dismantling

Every different type of job has a different union. You may have to pay an electrician, a carpet layer, a carpenter, a freight man, and a drayage man. (See why it gets expensive?) Prepare for this in advance. Most unions will let you take into your booth only what you can carry. And you can't even use your own dolly!

Logistics Checklist

_____ 1. Fill out freight storage stickers before you leave for the show.

_____ 2. Send all personnel a show map.

_____ 3. Distribute copies of flight schedules and hotel reservations.

_____ 4. Make copies of all bills of lading (shipping documents).

_____ 5. Have copies of all order forms you filled out and mailed to suppliers.

_____ 6. Check immediately upon arrival for any missing equipment.

_____ 7. Take phone numbers with you.

_____ 8. Lock your booth phone at night.

THE PRE-SHOW MEETING

Several days prior to the show, hold a pre-show meeting. Discuss any last-minute changes in plans, personnel, products, announcements, etc. The following is an example of a pre-show meeting agenda. Tailor it to your own needs, but at least include the following topics.

1. Goals for the show.
2. Products and services.
3. List of key customers to attend.
4. Show work schedule.
5. Coverage of any seminar sessions.
6. Selling practice.
7. Demonstration practice.
8. Qualification process.
9. Closing sequences.
10. Using the lead forms correctly.
11. Social hour.

DO'S AND DON'TS IN THE BOOTH

This is called boothmanship, probably borrowed from "salesmanship" years ago. Here's a final list of rules for your booth personnel to understand, memorize, and implement:

1. Dress, posture, and grooming are important.
2. Avoid attention-getting clothes.
3. Don't eat, drink, or smoke in the booth.
4. Don't huddle with colleagues; you are there to sell.
5. Don't sit down, unless your prospect sits first.
6. Wear your badge so that it's visible.
7. Try new opening lines, not "Can I help you?"
8. Use open-ended questions, like "Where might you be using our product?"
9. Don't discuss personal matters at the booth.
10. Don't prejudge people by their clothes.
11. *Remember, it's showtime!*

TRADE SHOW RESOURCES

Organizations

- Exhibit Designers & Producers Association, 1411 K Street, NW, #801, Washington, DC 20005.
- National Association of Exposition Managers, 334 S. Garfield Road, Aurora, CO 44202.
- Trade Show Bureau, 8 Beach Road, Box 797, East Orleans, MA 02643.

Books and Periodicals

- *How to Participate Profitably in Trade Shows,* Robert B. Konikow. Chicago: Dartnell Corp., 1985.
- *Exhibit Marketing,* Edward A. Chapman. New York: McGraw-Hill Books, 1987.
- *Tradeshow Week,* 12223 Olympic Boulevard, Los Angeles, CA 90064.

15

Networking: A Little Help from Your Friends

I've saved the best for last, because in the long run, friends, family, and colleagues are what it's all about. We can all use a little help now and then.

Reread the section on networking in Chapter 3. Then, if you need further people to contact (and who doesn't?), here are some final recommendations.

THE TEAM APPROACH

The invention process, from conception to marketplace, should be exciting, not fraught with fear. Part of the way to overcome the sense of aloneness is to build a solid team. The team approach is the foundation upon which all other elements of the business rest.

Here's who you should have on your team:

Your spouse, family, and friends
Business partners
Your lawyer

Accountants
Ad agencies
Bankers
Consultants
Manufacturer's representatives
Financial and business angels
College professors
Inventors' clubs
Entrepreneurs' clubs
Research and development facilities
Department of commerce
Small Business Administration
Better Business Bureau
Consulting engineers
Trade associations

ORGANIZATIONS FOR INVENTORS

AFFILIATED INVENTORS FOUNDATION, INC.
2132 E. Bijou Street
Colorado Springs, CO 80909
(800) 525-5885
Contact: Joanne M. Hayes
In business: 18 years

"To all inquiries, we provide a free Inventors Information Kit (including a Record of Invention form) and free consultation, if desired. We also perform evaluations and patent searches, publicity and marketing assistance services for minimum fixed fees which are kept low by high volume."

Fee: Two free preliminary appraisals per inventor per year; Patent search by a registered attorney, $275
Newsletter: Inventor's Digest (magazine), $20 per year

AMERICAN SOCIETY OF INVENTORS
P.O. Box 58426
Philadelphia, PA 19102-8426 (215) 546-6601
Contact: H. H. Skillman
In Business: 39 years

"Through member interaction, ASI counsels members and provides speakers in connection with making prototypes, doing searches; dealing with patent attorneys, invention brokers, venture capitalists and other entrepreneurs; evaluating inventions; and obtaining technical assistance."

Fee: $35.00 per year
Newsletter: Included with membership

CHICAGO HIGH TECH ASSOCIATION
211 W. Wacker Drive #1200
Chicago, IL 60606
(312) 939-5355
Contact: Sharon McCullough

"Fosters development and application of technology, enhances competition and economic growth of the region. Members include inventors, manufacturers, patent attorneys, venture capitalists, etc."

Fee: None
Newsletter: Quarterly.

IDEAS TO MARKET NETWORK
1320 High School Road
Sebastopol, CA 95472
(707) 829-2391
Contact: Norman C. Parrish
In Business: 3 years

"Ideas to Market Network publishes a newsletter 12 times a year specifically for the guidance of inventors and new business ventures. It is augmented by both technical and non-technical assistance on a one to one basis. The assistance is restricted to non-financial help. Telephone/fax service is free to members."

Fee: $50 per hour for consultation
Newsletter: $30 per year

INNOVATIVE PRODUCTS RESEARCH & SERVICES
P.O. Box 335
Lexington, MA 02173
(617) 862-5008
Contact: Donald D. Job
In Business: 6 years

"IPRS maintains a data bank of companies interested in new products and inventors with products, and facilitates connections. It works with manufacturers to teach them how to use inventions more effectively, and provides leadership in policy formulation to enhance U.S. competitiveness."

Fee: Hourly rates or percentage of gross
Newsletter: No

INVENTORS ASSOCIATION OF NEW ENGLAND
P.O. Box 335
Lexington, MA 02173
(617) 862-5008
Contact: Donald D. Job
In Business: 10 years

"Hold meetings, provide newsletters, workshops and networking of assorted resources. Annual Inventors Weekend for exhibiting inventions."

Fee: $25 per year
Newsletter: $25

INVENTORS' COUNCIL
53 W. Jackson Boulevard, Suite 1643
Chicago, IL 60604
(312) 939-3329
Contact: Don Moyer, President
In Business: 10 years

"Private non-profit educational corporation supported by contributions. Provides monthly workshops on how to get best patents and related issues. Makes cash awards."

Fee: None
Newsletter: Quarterly guides on topics

INVENTOR'S MARKETING ASSOCIATION
P.O. Box 2667
Chino, CA 91708-2667
(909) 393-8525
Contact: Reece Franklin
In Business: 4 years

"The purpose of Inventor's Marketing Association is to increase awareness of an inventor's idea to the various marketplaces he/she is targeting, through publicity, marketing, and networking. Quarterly publicity reports are sent to over 300 United States trade and consumer media, informing them of inventions now ready for market. Marketing seminars on advertising, publicity, and selling to manufacturers are held every three months."

Fee: $35 per year base fee; inclusion in quarterly reports starts at $75.00 per invention
Newsletter: Quarterly, included with membership

INVENTORS USA
132 Sterling Street
W. Boylston, MA 01583
(508) 835-3257
Contact: Barbara Wyatt
In Business: 14 years

"Programs monthly. Newsletter, support group, resource center."

Fee: $10 per year
Newsletter: Monthly, included with membership

INVENTORS WORKSHOP INTERNATIONAL
7332 Mason Avenue
Canoga Park, CA 91306-2822
(818) 340-4268
Contact: Alan Tratner
In Business: 22 years

"IWI is a non-profit, membership, volunteer organization that helps and guides the individual inventor in the development of ideas from inception to the market-ready stage by providing guidance analysis, idea protection services, Patent Saver™ program, books, seminars, expos, *Invent!* Magazine, networking."

> *Fee:* $35 for high school students; $75 for college students; $139 for full-time adult inventors (first year); $59 annual renewal
> *Newsletter:* Magazine, included with membership

MINNESOTA INVENTORS CONGRESS
P.O. Box 71
Redwood Falls, MN 56283
(507) 637-2344
Contact: Penny Becker
In Business: 36 years

"The Minnesota Inventors Congress, a non-profit organization, is dedicated to serving the inventor/entrepreneur through education, promotion, and referral. The MIC hosts the longest, continuous running Congress in the world, and provides year round service through our Inventors' Resource Center."

> *Fee:* Membership from $20 to $500
> *Newsletter:* Included with membership

OKLAHOMA INVENTORS CONGRESS
P.O. Box 54625
Oklahoma City, OK 73154-1625
(405) 848-1991
Contact: Albert N. Janco
In Business: 25 years

"The Oklahoma Inventors Congress is a non-profit organization with statewide chapters. Most chapters hold monthly meetings. There is an annual state meeting. Our members use their experiences to help new and potential members with problems. The speakers at our meetings are geared to help with current problems."

Fee: None
Newsletter: Included with membership

SOCIETY OF MINNESOTA INVENTORS
20231 Basalt Street
Anoka, MN 55303
(612) 753-2766
Contact: Paul G. Paris
In Business: 151 years

"To provide for ongoing education for its members in the innovation process. The Society is composed of inventors and others interested in helping inventors achieve success. To act as a liaison to other inventors' organizations. To facilitate the transfer of technology from the private inventor to the marketplace. To provide a forum for the discussion of problems common to private inventors."

Fee: $10 per year
Newsletter: Included with membership

TECHNOLOGY TRANSFER SOCIETY
611 North Capitol Avenue
Indianapolis, IN 46204
(317) 262-5022
Contact: Dr. Timothy Janis, Maureen Swinney

"As a society of professionals involved in the development, transfer, and commercialization of technology, we can link individuals and facilities that may assist them in development."

Fee: No charge for casual inquiry; full-service memberhsip, $75
Newsletter: Yes, also a quarterly journal and Directory of Members included with membership

WISCONSIN INNOVATION SERVICE CENTER
402 McCutchan
University of Wisconsin-Whitewater
Whitewater, WI 53190
(414) 472-1365
Contact: David Buchen
In Business: 12 years

"Identifying strengths and weaknesses of new products at an early stage can save developers a significant amount of time and money. For a nominal fee, WISC will perform preliminary market analyses of new product ideas. This aims to provide inventors with enough information to allow for improved decisions regarding further development."

Fee: $165 for evaluation services
Newsletter: No

NEWSLETTERS AND MAGAZINES

INTERNATIONAL NEW PRODUCT NEWSLETTER
Box 1146
Marblehead, MA 01945
(508) 750-4377
Contact: Pamela Michaelson
In Business: 35 years

"The newsletter lists over 80 products and processes available for license to and from all over the world, with contact names and addresses."

Fee: $150 per year

INVENT!
Mindsight Publishing
7332 Mason Avenue
Canoga Park, CA 91306-2822
(818) 340-4268
Contact: Alan Tratner

"*Invent!* is a bimonthly slick magazine for inventors and entrepreneurs. It is the magazine for Inventor's Workshop International, a non-profit organization with members and chapters nationwide."

Fee: Included with membership in Inventors Workshop International (see IWI listing).

INTERNATIONAL INVENTION REGISTER
c/o Catalyst
P.O. Box 547
Fallbrook, CA 92028
Contact: Dudley Rosborough

"Quarterly tabloid newspaper with worldwide offerings of patents, acquisitions, mergers, bulk transfers, financing, and industrial opportunities."

Fee: $18.00 per year

INVENTORS' DIGEST
c/o Affiliated Inventors Foundation
2132 E. Bijou Street
Colorado Springs, CO 80909-5950
Contact: John T. Faraday

"Published six times a year, slick magazine of the AIF. Includes local and national news, as well as success stories."

Fee: $15 per year

16

Packaging Your Invention

It should be obvious that *everything* you do in promoting and selling your product involves marketing. Your package mirrors that all-important company image.

Without a good package for your product, your product will not sell. Keep in mind that 90% of the reason for a product's success is due to packaging and marketing and only 10% to the product itself. My late friend Ric Rasmussen, one of the founding members of Los Angeles' famous Magic Castle, the international Magicians Association said, "90% of a magic trick is show, and only 10% is magic." So your package must have sales appeal—the sexy sizzle.

INTEGRATING PACKAGING IN YOUR MARKETING PLAN

It is vital that you integrate your packaging into your overall marketing plan. Your packaging should reflect the theme of your marketing campaign, and it must appeal to your target market. If you have more than one market you intend to sell to, you may need more than one package.

We can boil down packaging into four simple topics for consideration:

1. You must know who your primary and secondary target markets are. (Use the basics of market research—age, sex, race, etc.—to determine your primary market.)
2. You must design your package to appeal to your primary market first.
3. You must consider all ethnic and regional factors, but don't integrate them until your primary audience has been satisfied.
4. You must consider how you intend to sell the product. Will it be by mail order, on retail shelves, direct to consumers through party plans, or what?

Thus, one step at a time. Now consider some of the different markets that your product could be targeting.

THE TARGETED APPROACH

In Chapter 11, we discussed how to target your market, how to research their needs and wants. To recap: There are hundreds of different markets you *could* choose from. For example:

- Teenagers
- Yuppies (young urban professionals)
- Dinks (double-income, no kids)
- Mature (seniors)
- Couples
- Ethnic groups (race and religion)
- Regional groups

You must analyze which ones will be the most profitable for you, based on what your product is used for. You obviously had a primary market in mind when you first came up with the idea, right? Take that market, and subdivide it into buying segments, such as those shown above.

Now, do some research. (The Library of Congress, for example, has lots of marketing surveys on various subjects you can get inexpensively. Contact them in Washington, DC, tell them the subject you want to research, and ask if any major surveys have been done on the subject.)

Your research should show you how much your potential markets are buying. Then check out some trade publications, and send for their annual surveys. These "Year in Review" issues will tell you whether the market is going up or down, and the amount of money spent on a particular industry's products by market.

When you've determined whether or not you're on track with your primary market, it's time to analyze how it responds to different stimuli.

FIVE KEY RULES IN PACKAGING

1. Good visual appeal is the first of the five key rules in good marketing packaging. The product must be visually appealing, or people will pass it by. This principle applies to *any* and *every* selling method you use.

2. Consider whether your product is right for POP, point-of-purchase. POP packaging is placing a stack of your products right next to the cash register, encouraging customers to make an "impulse buy." You want the retailer to feel your invention will sell like hotcakes, so he'll place your products for POP.

3. Your package should be lightweight. It may look good, but if it's too heavy, customers may change their mind about buying it.

4. Your products should be easy to shelve or stack, for both the store and the consumer.

5. The package itself must sell. It must have psychological selling colors, and it must be the right shape and size.

VISUAL MESSAGES IN PACKAGING

The different visual messages in packaging include color, size, and shape. Different colors have different psychological meanings, and shape and size will make a package easy or difficult to handle and store.

Color

The primary colors (remember your first-grade art class?) are red, blue, and yellow. Add black (the entire color spectrum in one) or white (the absence of color), and you have a rainbow.

Printing Colors. As you know, your packaging will have colors printed on it. In printer's terms, the primary colors are process blue, process red, and process yellow. If you ask for two colors, the printer will give you black plus one process color. Therefore, green is not two colors, it's three—black (always counted by the printers), blue (cyan), and yellow.

The exception to the above rule is when you choose a PMS color. PMS colors are premixed to certain specifications. You can choose a PMS red, for example, off a chart, which is considered two colors (black plus the PMS color). Sometimes, printers will only charge you the two-color charge for PMS, whereas others will charge for three colors. Be sure to check with your printer.

The Psychology of Colors. The colors you choose for your packaging should be determined by two considerations: what looks pleasing to the consumer (i.e., what will sell well), and what psychological mood you wish to create.

If your product is a powerful one, you might choose red. Red says power and is a strong selling color. Yellow is a bright, lighter sales color. If your product denotes calm and quiet, you might choose a greenish tint (not dark or kelly green, but soft like aqua). A stronger green, with a tint of yellow in it, would convey a refreshing sensation. The color orange connotes an action product. Blue is cool, calm, and serene. And, of course, pink is feminine, while all black, like the packaging for Drakkar™ men's cologne, implies masculine and powerful.

Shape

The shape of your product's package has a lot to do with how well it will sell. For example, cologne for men usually comes in a package that says sexy and macho. Women's packages often denote femininity and sometimes have a sexy aspect as well.

You don't have to be blatant. You can sell your product using good colors, without resorting to vulgar packaging.

Children's packaging usually indicates fun, roughhousing, playing. The colors are often bright and alive, saying mischievous.

Packaging for teenagers is trendy, upbeat, and sometimes faddish. The colors are bright and lively, like a wild painter's palette.

Packaging for mature customers is calm and soothing, with solid-as-a-rock style.

Size

The size of your packages, both internal and external, will be determined by what the product is and how you intend to display it.

The larger a package is, the more shelf space and room it takes up. Consider whether you will overburden the customer in terms of storage in the home. When dealing with the retailer, don't forget that he has many products to display, and shelf space is probably tight. Try to make it easy for him to accept your product. The lighter, the smaller, the better it looks, the more likely you will get shelf space in stores. (Some stores, like grocery chains, even charge you a fee—known as a "slotting fee"—for putting your product on the shelves. They range from $500 to $15,000, so beware!)

THE FOUR TYPES OF DISPLAYS

You must be aware that in a retail environment, there are four basic places to display a product: shelf, counter, wall, and floor. Your package should ideally be created so you can take advantage of all four locations. You may not know how a retailer will feel about your product, or how much shelf space he is willing to give up.

HOW SHOULD YOU PACKAGE IT?

There are four main ways to package a product: blister pack, shrink-wrap, plastic bag, and box.

A *blister pack* is used for razor blades, batteries, and similar products. A hard plastic shell is glued to the front of a printed cardboard backing, with the product in the middle. The plastic conforms its shape to the product. This can be a very expensive process. Normally, small items are blister packed. (Remember, there's a premium on store space.)

We're all familiar with *shrink-wrapping*. Some uses are audiotapes and videotapes. Shrink-wrapping implies freshness, a brand-new whatever just for you! Shrink-wrapping is relatively inexpensive, but you still need a cardboard backing or box for support.

Plastic bags are inexpensive packages for your product. They can be silk-screened or overprinted, and they come in many sizes. The problem with plastic bags is that they look cheap, and you probably don't want to send that message.

Most products with medium to high cost are put into *boxes*. This is really the best way to go. The entire box is another sales message to the customer.

The front should include an *action* picture of the product, the name, a headline, and a few good, strong points of selling copy. The end panels include the name and the size or quantity of the product, and the back panel should include an action photo or a drawing of the product in use. Also included on the back would be copy saying why the customer needs the product, and a brief summary of use or assembly instructions. (Detailed user instructions will be inside the box.)

NAMING YOUR PRODUCT

While we talked about naming your company in Chapter 5, let's address the *product name* here. You must name your product with salability in mind. "Whizbang" won't cut it, unless you have enough money to do a saturation advertising and promotion campaign, so that *everyone* knows what a Whizbang is. And face it, you don't have that kind of money.

I suggest you take the "kis" approach: Keep it simple. The simplest way is to make sure the name describes what the product is, or what it does. For example, Handi Wipes tells what the product does and how simple it is to use. On the other hand, Bounty paper towels had to spend a lot of money to convince the consumer that their towels were better. And, they have to add the words "paper towels" each time the name comes up.

Try to take the Handi Wipe approach, if you can.

SOME FINAL THOUGHTS

While this chapter has given you a brief glimpse into how packaging integrates with marketing, it is by no means definitive. Check your library's card catalogue or computer under "packaging—industrial."

Finally, there are consultants who specialize in the packaging aspect of marketing. Check your business-to-business yellow pages, and the American Marketing Association in Chicago for names of packaging consultants in your area. A few dollars spent will probably increase the salability of your product tenfold.

APPENDIX

PUBLICITY FORMS

These publicity forms are examples to be used with Chapter 12. They are samples only, and may differ from the final copy you write. Use them as brain teasers, or examples of how to word and format specific publicity tools. The final wording should be your own. Make them unique, using your own style of writing. This will win an editor's heart. In other words, don't copy verbatim.

Appendix: Publicity Forms 243

PUBLICITY PROJECT TRACKING SHEET

PROJECT: _____ DUE DATE: _____

GOALS SET _____

    WHOM DO WE WANT TO REACH?

    WHY DO WE WANT TO REACH THEM?

MEDIA ANALYSIS

    THEY MIGHT READ:

    THEY MIGHT WATCH:

    THEY MIGHT LISTEN TO:

APPROACH:

    GEOGRAPHIC _____

    DEMOGRAPHIC _____

RESEARCH:

    BACON'S _____

    WORKING PRESS _____

    GALE'S _____

    SRDS _____

    RGPL _____

SELECTION:

    MAGAZINES _____

    NEWSPAPERS _____

    RADIO _____

    TV _____

OBTAIN SAMPLES:

    MAGAZINES/NEWSPAPERS/RADIO/TELEVISION–3 OF EACH;
    FOCUS ON:

    STYLE _____

        PHILOSOPHY _____

        REPORTERS _____

WRITE IT ALL DOWN

PREPARE THE NEWS RELEASE

SEND THE NEWS RELEASE

TRACK THE RESULTS

Appendix: Publicity Forms 245

(Media Advisory Format)

(On Your Letterhead)

Name of Event:

Day/Date:

Time:

Place:

Contact:

Business Phone:

HEADLINE

1st Paragraph:

2nd Paragraph:

3rd Paragraph:

4th Paragraph:

ACTION STEP:

(News Release Format)

(On Your Letterhead)

FOR IMMEDIATE RELEASE

For Further Information,
Contact: John Doe
(312) 123-4567

HEADLINE

(Dateline) — Begin paragraph 1 here . . .

Paragraph 2:

Paragraph 3:

Paragraph 4:

Paragraph 5:

ACTION STEP:

Appendix: Publicity Forms 247

(Public Service Announcement)

(On Your Letterhead)

Public Service Announcement

Title

Timing/Word Count

For Use Beginning: Date

ANNCR:

(Company Backgrounder Format)

(On Your Letterhead)

COMPANY BACKGROUNDER

NAME:

PURPOSE:

OFFICER #1:

OFFICER #2:

OFFICER #3:

OFFICER #4:

ADDRESS:

PHONE:

CONTACT:

(Fact Sheet Format)

(On Your Letterhead)

FACT SHEET

PRODUCT NAME

SPECIFICATIONS:

DESCRIPTION:

MARKETS:

AVAILABILITY:

CONTACT:

PHONE:

MEDIA ANALYSIS GRID

☐ **Radio**   ☐ **TV**   ☐ **Newspaper**   ☐ **Magazine**

What They Accept	Media Vehicle #1	Media Vehicle #2	Media Vehicle #3
Publication Date/Day			
Subject Editor/ Subject Reporter			
Releases			
Photos			
Feature Articles			
Op-Ed Pieces			
Letters to Editor			
Media Advisory			
PSAs			
Film			
Videotape			
Tape Recordings			
Freelance Copy			
New Product News			
Personnel News			
Q and A			
Calendar Listings			

Appendix: Publicity Forms 251

MEDIA TRACKING SHEET

MEDIA   DATE SENT   TO WHOM SENT   USED   INCHES/TIME

Afterword

As I said in the dedication at the beginning of the book, in their zest to create, inventors shine above all others. I really mean that!

My relationship with inventors goes back some twenty years, to the time I met the inventor of the Lava Lite in Chicago. My mother was his secretary for a short time, and I found his tinkering fascinating. I have always felt that inventors are a special breed. I know you think so, too.

The creation of any invention takes time and nurturing. My latest "invention" is the book you have just finished. Like your projects, it has been a labor of love.

I have done my best to give you insightful information about the different processes that are involved in invention marketing. What you choose to do with this information is up to you.

I wish you luck in your goals and dreams. Perhaps we'll meet at an inventors' convention or workshop. Until we do, keep on target, and remember: *Don't let anyone burst your dream machine!*

Bibliography

Chapter 2: The Skills You Need

Barnhart, Helene Schellenberg. *How to Write & Sell the 8 Easiest Article Types.* Cincinnati, OH: Writer's Digest Books, 1985.

Cathcart, Jim. *Relationship Selling: How to Get and Keep Customers,* 1992. J. C. Publications, P.O. Box 9075, La Jolla, CA 92038. (800) 222-4883.

Dawson, Roger. *The Secrets of Power Negotiating.* Lincolnwood, IL: Nightingale-Conant, 1987. Audiotape.

Fisher, Roger, and Ury, William L. *Getting to Yes.* New York: Viking Penguin, 1991.

Fletcher, Leon. *How to Speak Like a Pro,* New York: Ballantine, 1985.

French, Christopher W., ed. *The Associated Press Stylebook and Libel Manual.* Reading, MA: Addison-Wesley, 1987.

Goman, Carol Kinsey. *Creativity in Business.* Los Altos, CA: Crisp Publications, 1989.

Hopkins, Tom. *How to Master the Art of Selling Anything.* New York: Warner Books, 1988.

Strunk, William, and White, E. B. *Elements of Style.* New York: Macmillan, 1979.

U.S. Small Business Administration. *Checklist for Going into Business.* Management Aid 2.016. Washington, DC: U.S. Government Printing Office, 1989.

Winston, Stephanie. *The Organized Executive.* New York: Warner Books, 1985.

Chapter 3: Getting Started Right

Burdek, Deborah M., ed. *Encyclopedia of Associations.* Detroit: Gale Research, 1993.

Business Plan for Small Manufacturers. Management Aid 2.007. Washington, DC: U.S. Government Printing Office, 1987.

Jenkins, Michael. *Starting and Operating a Business in California.* Grants Pass, OR: Oasis Press, 1993.

Mancuso, Joseph R. *How to Write a Winning Business Plan.* New York: Prentice-Hall Press, 1985.

Olmi, Antonio M. *Selecting the Legal Structure of Your Firm.* Management Aid 6.004. Washington, DC: U.S. Government Printing Office, 1989.

Stetler, Susan L., ed. *Brands and Their Companies.* Detroit: Gale Research, 1993. Formerly titled *Trade Names Directory.*

Chapter 4: From Ideas to Product
Martin, Charles L. *Starting Your New Business.* Los Altos, CA: Crisp Publications, 1988.

U.S. Small Business Administration. *Finding a New Product for Your Company.* Management Aid 2.006. Washington, DC: U.S. Government Printing Office, 1989.

Chapter 6: Financing Your Invention
Blum, Laurie. *Free Money for Small Business and Entrepreneurs.* New York: Wiley, 1992.

Davis, Steven, ed. *Writer's Yellow Pages.* Dallas, TX: Steve Davis Publishing, 1988.

Foundation Directory, Detroit: Gale Research, 1993.

Lesko, Matthew. *Government Giveaways for Entrepreneurs.* Chevy Chase, MD: Info USA, 1989.

Polk's Bank Directory. Nashville, TN: R. L. Polk, 1992.

Pratt, Stanley. *Guide to Venture Capital Sources.* Boston: Venture Economics Inc., 1987.

Renz, Loren, and Olson, Stan, eds. *The Foundation Directory.* New York: Foundation Center, 1987.

Chapter 7: Budgeting and Scheduling
Chase, William D., and Chase, Helen M. *Chase's Annual Events.* Chicago: Contemporary Books, 1993.

Mancuso, Joseph R. *Mancuso's Small Business Resource Guide,* New York: Prentice-Hall Press, 1988.

Ries, Al, and Trout, Jack. *Positioning: The Battle for Your Mind.* New York: Warner Books, 1987.

Valdez, Gene, and Frost, Charles, *How to Prepare a Bank Financing Proposal for Your Business: The Way a Banker Would,* Montclair, CA: 1992.

Chapter 8: Selling Out for Big Profits
Dun & Bradstreet Middle Market Directory. Parsippany, NJ: Dun's Marketing Services, 1992.

Dun & Bradstreet Million Dollar Directory. Parsippany, NJ: Dun's Marketing Services, 1992.

Levy, Dick. *Inventor's Desktop Companion,* Detroit: Visible Ink Press, 1991.

Standard & Poor's Corporation Records. 7 Vols. New York: Standard & Poor's, 1992.

Thomas Register of American Manufacturers. 21 Vols. 82nd Ed. New York: Thomas Publishing, 1992.

Chapter 9: Distribution

Direct Marketing Market Place. Boca Raton, FL: Hilary House, 1989.

Directory of Department Stores. New York: CSG, Lebhar-Friedman, 1992.

Directory of Discount Stores. New York: CSG, Lebhar-Friedman, 1992.

Directory of General Merchandise. New York: CSG, Lebhar-Friedman, 1992.

Directory of Manufacturers' Sales Agencies. Laguna Hills, CA: Manufacturers' Agents National Association, 1989.

Fairchild's Financial Manual of Retailers. New York: Fairchild, 1992.

Gottlieb, Richard. *Directory of Mail Order Catalogs.* Lakeville, CT: Grey House, 1989.

Graeser, Kathi, ed. *Electronic Representative's Directory.* Twinsburg, OH: Harris Publishing, 1987.

Klein, Bernard. *Mail Order Business Directory.* West Nyack, NY: Todd Publications, 1992.

Phelon's Directory. Fairview, NJ: Phelon Sheldon & Marsar, 1990.

Chapter 10: Selling Your Invention to the U.S. Government

U.S. Department of Commerce. *Commerce Business Daily.* Washington, DC: U.S. Government Printing Office, 1993.

U.S. Small Business Administration, *U.S. Government Purchasing and Sales Directory.* Washington, DC: U.S. Government Printing Office, 1990.

Chapter 11: Marketing

Martin, Charles L. *Starting Your New Business.* Los Altos, CA: Crisp Publications, 1988.

Chapter 12: How to Get Free Publicity

All in One Directory. New York: Gebbie Press, 1993.

Bacon's Publicity Checker. 2 Vols. Chicago: Bacon's Publishing, 1993.

Boyden, Donald P., and Krol, John, eds. *Gale Directory of Publications.* Detroit: Gale Research, 1990.

Broadcasting/Cablecasting Yearbook. Washington, DC: Broadcasting Publications, 1993.

Burdek, Deborah M., ed. *Encyclopedia of Associations.* Detroit: Gale Research, 1993.

Burgett, Gordon. *How to Set Up and Market Your Own Seminar.* Santa Maria, CA: Communications Unlimited, 1987.

Franklin, Reece A. *101 Ideas for News Releases.* Chino, CA: AAJA Publishing, 1989.

Magazine Rates and Data. Wilmette, IL: Standard Rate and Data Service, 1993.

Manning, Matthew, ed. *Oxbridge Directory of Newsletters.* New York: Oxbridge Communications, 1993.

Neff, Glenda T., ed. *1993 Writer's Market.* Cincinnati, OH: F & W Publications, 1993.

Newspaper Rates and Data. Wilmette, IL: Standard Rate and Data Service, 1993.

Shepard, Kathy, ed. *Southern California Media Directory.* Los Angeles: Publicity Club of Los Angeles, 1993.

Spot Radio Rates and Data. Wilmette, IL: Standard Rate and Data Service, 1993.

Spot Television Rates and Data. Wilmette, IL: Standard Rate and Data Service, 1993.

Towell, Julie E., ed. *Directories in Print.* 2 Vols. 6th ed. Detroit: Gale Research, 1993.

Working Press of the Nation. 5 Vols. Chicago: National Research Bureau, 1993.

Chapter 13: Advertising

Betancourt, Hal. *The Advertising Answerbook.* Englewood Cliffs, NJ: Prentice-Hall, 1982.

Broadcasting/Cablecasting Yearbook. Washington, DC: Broadcasting Publications, 1993.

Caples, John. *How to Make Your Advertising Make Money.* Englewood Cliffs, NJ: Prentice-Hall, 1983.

Cohen, William A. *Building a Mail Order Business.* New York: Wiley, 1991.

Davis, Steven, ed. *Writer's Yellow Pages.* Dallas, TX: Steve Davis Publishing, 1988.

Franklin, Reece A. *How to Advertise and Promote Your Business.* Chino, CA: AAJA Publishing, 1993.

Goodman, Gary S. *You Can Sell Anything by Telephone.* Englewood Cliffs, NJ: Prentice-Hall, 1984.

Levinson, Jay Conrad. *Guerrilla Marketing.* Boston: Houghton Mifflin, 1984.

Levinson, Jay Conrad. *Guerrilla Marketing Weapons.* New York: Plume, 1990.

Lorenzen, Jim. *Ad Strategies That Work.* Orlando, FL: Lorenzen Associates, 1983. Audiotape.

Magazine Rates and Data. Wilmette, IL: Standard Rate and Data Service, 1993.

Newspaper Rates and Data. Wilmette, IL: Standard Rate and Data Service, 1993.

Roman, Kenneth, and Maas, Jane. *How to Advertise.* 2d ed. New York: St. Martin's Press, 1992.

Schwab, Victor O. *How to Write a Good Advertisement,* Los Angeles: Wilshire Book Company, 1980.

Spot Radio Rates and Data. Wilmette, IL: Standard Rate and Data Service, 1993.

Spot Television Rates and Data. Wilmette, IL: Standard Rate and Data Service, 1993.

Watkins, Don. *Newspaper Advertising Handbook.* Wheaton, IL: Dynamo, 1984.

Chapter 14: Trade Show Secrets of the Pros

Chapman, Edward A. *Exhibit Marketing.* New York: McGraw-Hill, 1987.

Konikow, Robert B. *How to Participate Profitably in Trade Shows.* Chicago: Dartnell, 1985.

Trade Show Data Book. Los Angeles: Trade Show Week Publications, 1993.

Trade Show Week. Los Angeles: Trade Show Week Publications, 1993.

Chapter 15: Networking

Faraday, John T. *Inventor's Digest.* Colorado Springs: Affiliated Inventors Foundation, 1993.

Michaelson, Pamela. *International New Products Newsletter.* Marblehead, MA: Michaelson Publications, 1993.

Rosborough, Dudley. *International Invention Register.* Fallbrook, CA: Catalyst, 1993.

Tratner, Alan A. *Invent Magazine.* Canoga Park, CA: Mindsight Publications, 1993.

Index

A

Accountants, financing invention with, 72
Accounts payable, 12
Accounts receivable, 12
Action notice on patents, 61-62
Ad Strategy that Works (Lorenzen), 179-181
Advertising, 161-208. *See also* Direct-mail package; Magazines; Newspapers; Radio; Television
   bids, U.S. government advertising for, 112
   in broadcast media, 166-168
   chart of advertising methods, 162
   classified ads, 97-98, 187
   co-op advertising, 183
   direct mail advertising, 162-164
   display advertising, 186-187
   fictitious business name, 27
   finding right media for, 200-203
   layouts for, 203-204
   mail-order advertising, 204-208
   media cost-comparison grid, 202
   in print media, 164-166
   resources for, 208
   sample ad, 98
   selling your invention with, 97-98
   on signs, 169
   specialties, 169
   trade show promotion through, 220
Advertising agencies, marketing by, 133-134
Advertising Answerbook, The (Betancourt), 208
Advertising Club Referral Service, 134
ADWEEK, 134, 194
Affiliated Inventors Foundation, Inc., 226
Agencies for marketing, 133-134
   interview questions for, 134-137
   payment schedule for, 136-137
Agriculture, loans through U.S. Department of, 77
AIDA selling technique, 15
   formula for direct-mail package, 174-175
Air Force, department of, 118
Amendments of patents, 62
American Marketing Association, 240
American Society of Inventors, 227
AM radio, 196
Annual Directory of Events (Chase's), 88
Appealing denial of trademark application, 55
Apple computers, 7
Appliances, 6
Army, Department of, 118

259

Artist's rendition of idea, 46
Assignment of patents, 63
Associated Press Stylebook, The, 14
Associations and clubs. *See also* Community organizations; Industry associations
  directories for, 147
  for inventors, 226–233
  leads club, 17
  speaking skills and community organizations, 15
  writers' club, 14
Attorneys. *See also* Patent attorneys
  protecting invention, use in, 49–50
  trademark search by, 55
Audio cassettes on negotiating skills, 18
Automobile insurance, 30

B

Bacon's Publicity Checker, 145–146
Balance sheet, current, 34–35
Banks
  financing invention through, 72–73
  ideas from, 41
Barnhart, Helene Schellenberg, 14
Baylor University, 100
Betancourt, Hal, 208
Bidders' lists, 112
  form of solicitation mailing list application, 116
  GSA bid list, 113
  procedure for getting on, 115–116
Big goals, 22
Billboard advertising, 169
Black media, 145
Blister pack packaging, 239
Blum, Laurie, 74

Bookkeeping skills, 12–13
Books. *See also* Copyrights
  on negotiating skills, 18
  on selling, 15–16
  on speaking skills, 15
  on writing skills, 14
Booth at trade shows, 214–218
Boston, 190
Boston Globe, 165
Boxes, packaging in, 239
Boy Scout directories, 165–166
Brands and Their Companies (Gale's), 28, 55
Break-even analysis, 83–84
Brennan, Gary, 6
Broadcasting/Cablecasting Yearbook, 147, 166
Brochures, 164
Bubble packs, 6
Budgeting. *See also* Business records
  all you can afford budgeting, 86
  break-even analysis, 83–84
  last year + X% budgeting, 86
  for magazine advertising, 192–193
  for marketing, 85–87
  match the competition budgeting, 86
  monthly budget guideline chart, 88
  objective-and-task method budgeting, 86–87
  percentage of sales budgeting, 86
  for public relations, 150
  rates for advertising in, 193–194
  reasons for, 81–82
  simple method for, 87
  trade show budgets, 212–214
  types of records for, 82–84

updating records, 82
Building a Mail Order Business (Cohen), 181
Burgett, Gordon, 151
Business development corporations, 73
Businesses
  financial statement for, 32–35
  insurance for, 30–31
  records, types of, 82–84
  types of, 23–25
Businesses Turned Down Elsewhere, SBA Loans for, 77
Business magazines, 189–190
Business plan
  financing invention, use in, 70–71
  formulation of, 35–36
Business/Towns Less Than 50,000 People, USDA, 77

C

Cabinetmakers, prototypes by, 46
Cable television, 168
  versus regular TV, 196–197
  tips for good commercials, 199
California Business, 190
Caples, John, 208
Card catalog of library, 149
Cash flow, 12
  forecast, 34
  tracking of, 83
Cash receipts, 83
Catalog of Federal Domestic Assistance, 77
Catalogs. *See also Directory of Mail Order Catalogs*
  advertising in, 162
  manufacturers' catalogs, 44
Cathcart, Jim, 16

Center for Entrepreneurial Management, The (CEM), 36
Certificate of Competency program, SBA, 117
Chain stores, directory of, 108
Chambers of Commerce, 15
  special event scheduling and, 88
Chapman, Edward A., 224
Charity directories, 165–166
Chase's *Annual Directory of Events*, 88
Check registers, 83
Chicago High Tech Association, 227
Chicago Tribune, 165
Children's packaging, 238
Church bulletins, advertising in, 165–166
Circulars, 164
Civic groups. *See* Community organizations
Civilian government agencies, sales to, 119
Class CA copyright, 64
Classified ads, 97–98, 187
  for mail-order advertising, 207
Classified display advertising, 186–187
Classified throwaways, 190
Class RE copyright, 64
Class TX copyright, 64
Class VA copyright, 64
Clip sheets
  electronic clip sheets, 156–157
  print clip sheets, 155–156
Closing the Sale, 15
Cohen, William, 181
Colors
  newspaper ads, 189
  for packaging of invention, 237
Combination type of invention, 4, 6

Commerce Business Daily, 113
Commissioner of Patents and Trademarks, 57
Community college classes
  speaking skills and, 15
  for writing skills, 14
Community organizations
  networking through, 17–18
  speaking skills and, 15
Company backgrounder, 153
  project tracking sheet for, 248
Competition
  analysis grid, 128
  evaluation checklist, 131
  and marketing, 127–128
Computer
  industry, 7
  Microcomputer Marketplace, 109–110
  writing software, 175
Conferences
  press kit at, 158
  and publicity, 151
Conscious effort for invention, 7
Consumer magazines, 190
Co-op advertising, 183
Copyright Office, 66
Copyrights
  agency granting, 52
  classes of, 64
  defined, 63
  duration of, 66
  form of copyright application, 65
  length of time to process, 64
  registration of, 63
  self-registration of copyright, 52
Corporations, 24–25
  financing invention with, 75
  ideas from, 41
Counter displays, 238
Coupon books, 163

Cover letter with news release, 149–150
Creative skills, 9–10
Creativity in Business (Goman), 9–10
Creativity quotient (CQ), 9–10
Credit unions, 73
Current balance sheet, 34–35
Customer potential, 40

D
Dailies, advertising in, 165
Daily problem-solving ideas, 7
Dawson, Roger, 18
Day Planner, 16
Decision-making skills, 12
Decline stage for sales, 103
Defense Logistics Agency (DLA), 113
Department stores, directory of, 107
Descriptions
  of business, 36
  selling your invention with, 97
Direct effort to invent, 7
Direct-mail package, 162–164
  AIDA formula for, 274–275
  brochure, 170–171, 172–173
  compared to mail-order advertising, 204
  credit terms in, 180–181
  keys to success with, 179–181
  order form for, 176–178
  outer mailing envelope, 178–179
  response card, 177–178
  return envelope, 178
  sales letter for, 171, 174–175
  self-mailers with, 176
  tear off sheet/card, 178
  writing tips, 175

Direct Marketing Market Place, 109–110
Directory of Department Stores, 107
Directory of Discount Stores, 107–108
Directory of Drug Stores, 109–110
Directory of General Merchandise, Variety Chains, and Specialty Stores, 108
Directory of Mail Order Catalogs, 106
for specialities, 109
Directory of Manufacturer's Sales Agencies, 105
Directory of Registered Patent Attorneys and Agents, 60
Disclosure Document Program for patent, 59
Disclosure statements. *See also* Nondisclosure agreement
selling invention and, 95–96
Discounts, 85
newspaper discount rates, 187–188
Discount stores, directory of, 107–108
Discovery Channel, 196, 197
Disney Channel, 196
Display advertising, 186–187
Distribution, 102–110
flowchart, 104
manufacturer's reps for, 105
normal sales cycle, 103
Distributors, 104–105
master distributors, 105
Drawings, selling your invention with, 97
Drug stores, directory of, 109–110
Dun & Bradstreet Middle Market Directory, 95
Dun & Bradstreet Million Dollar Directory, 95

E
Electronic Representative's Directory, 105
Elements of Style, The (Strunk & White), 14
Elks Clubs, 151
Empty niche, capitalizing on, 127, 129
Encyclopedia of Associations regional editions, 151
Encyclopedia of Associations (Gale's), 28, 147
Energy Loans for Inventors (SBA), 77
Energy technology
Energy Loans for Inventors (SBA), 77
Office of Planning and Coordination, 79
Entrepreneur, 189
Entrepreneurial skills, 4, 10–13
Envelopes, 29–30
with direct-mail package, 178–179
Environmental impact, evaluation of, 43
ESPN, 196
cost of advertising on, 197
Estimated cash forecast, 34
Evaluation of invention, 99–100
Exciting goals, 22
Exclusive patent license, 101
Exhibit booth at trade shows, 214–218
Exhibit Designers & Producers Association, 224
Exhibit Marketing (Chapman), 224
Expected sales and expenses, 33, 34

F
Fact sheets, 153–154
project tracking sheet for, 249

Fairchild's Financial Manual of Retailers, 108
Family. *See also* Networking
financing through, 73
Farming out
manufacture of invention, 94–95
marketing, 132–134
Favoritism in media, 139
Feature articles, 154
Federal Communication Commission (FCC) licenses, 28
Federal government. *See* United States government
Federal Procurement Data Center (FPDC), 113
Federal Property Resources Service (FPRS) of GSA, 115
Federal Supply Service (FSS) of GSA, 115
Fees
for copyright registration, 63
for Disclosure Document Program for patent, 59
evaluating inventions, 99–100
for fictitious business name permits, 27
for marketing agencies, 136–137
for patents, 62
trademark fees, 29
for trademarks, 55, 57
Fictitious business name permits, 26–27
Financial statements
business financial statement, 32–35
personal financial statement, 31–32
Financial structure, determination of, 31–35
Financing invention, 69–70
accountants for, 72
banks, 72–73

Financing invention (*continued*)
  business development corporations, 73
  business plan, use of, 70–71
  credit unions, 73
  family and friends, 73
  foundation and grant financing, 73–74
  government programs for, 76–77
  incubators for, 75
  Information Officer, 79–80
  insurance policies for, 75
  inventors' clubs and, 75
  manufacturers and corporations for, 75
  NBS Office of Energy-Related Inventions, 79
  Office of Planning and Coordination, 79
  personal funds, financing with, 75–76
  presentation to lender, requirements for, 70
  real estate equity loan financing, 76
  savings and loan associations, 72–73
  small business investment companies (SBIC), 77
  sources of funding, 71–72
  stock placement, financing with, 76
  terminology of, 78
  venture capital clubs for, 78
  Xerox Foundation, 79
Finding a New Product for Your Company (SBA), 41
Fire insurance, 30
Fisher, Roger, 18
Fixed spot/rate, 197
Fletcher, Leon, 15
Floaters, insurance, 30

Floating spot, 197–198
Floor displays, 238
Flowchart
  distribution, 104
  for getting started, 21
  for marketing invention, 5
Flyers, 164
FM radio, 196
Focus group, test-marketing to, 46–47
Food and Drug Administration (FDA) licenses, 28
Foreign license, ideas from, 41
Foundation Center Directory, 74
Foundation Directory, 74
Foundations
  financing through, 73–74
  Xerox Foundation, 79
FPRS of GSA, 115
Franklin, Benjamin, 7
Free Money for Small Business and Entrepreneurs (Blum), 74
Freight costs, 85
Friends. *See also* Networking
  financing through, 73
FSS of GSA, 115
FTC Mail-Order Rule, 204
Full-facility wholesalers, 103–104

G
Gale Research
  Brands and Their Companies, 28, 55
  Encyclopedia of Associations, 28
  Gale Research's Directories in Print, 145
  Newsletter Directory, 147
Gale's Directory of Publications, 146
  for clip sheets, 155

Gang run printing, 29–30
Gebbies All-in-One Press Directory, 146
General business insurance, 30
General Information Concerning Patents, 63
General journal, 83
General ledger accounts, 12
General Services Administration (GSA)
  Business Service Centers of, 113
  purchasing programs of, 115
Geographic considerations
  business magazines, 190
  for marketing, 126
Getting to Yes (Fisher), 18
Girl Scout directories, 165–166
Goal-setting, 21–22
Goman, Carol Kinsey, 9–10
Goodman, Gary, 170
Government Giveaways for Entrepreneurs (Lesko), 76
Government programs and organizations. *See also* Small Business Administration (SBA); United States government
  financing through, 76–77
  local government financing, 76, 77
  selling product to, 106
Gramattik VI, 175
Grantmanship Center, 74
Grants
  program financing, 73–74
  proposals, 74
Grocer's Marketing Guide Book, 109–110
GSA bid list, 113
Guerrilla Marketing (Levinson), 185

Guide to Venture Capital Sources (Pratt), 78

H
Handbills, 169
Health insurance, 30
Health issues, 12
Health permits, requirements for, 28
Highlander chain, 165
Hispanic media, 145
Holidays, scheduling and, 88
Hopkins, Tom, 16
How to Advertise (Roman & Maas), 208
How to Make Your Advertising Make Money (Caples), 208
How to Master the Art of Selling Anything (Hopkins), 16
How to Participate Profitably in Trade Shows (Konikow), 224
How to Prepare a Bank Financing Proposal for Your Business (Valdez), 78
How to Set up and Market Your Own Seminar (Burgett), 151
How to Speak Like a Pro (Fletcher), 15
How to Write and Sell the 8 Easiest Article Types (Barnhart), 14
How to Write a Winning Business Plan (Mancuso), 36
Human interest and publicity, 140

I
Idea Book
 creativity and, 10
 protecting invention with, 50
Ideal products, 40
Ideas, 6–7. *See also* Products
 defined, 39–40
 evaluation of, 43
 feasibility study on, 44–45
 looking for, 41
 mind map for generation of, 43
 newness, checking for, 44
 patents on, 61
 tough questions about, 42–43
Ideas to Market Network, 227–228
Improving your invention, 50
Inc. Magazine, 189
Incubation stage for sales, 103
Incubators, financing through, 75
Incubator Times, 75
Independent Bankers Association, 72
Independent contractors for marketing, 132–133
Industry associations, 17
 ideas from, 41
Informal protection, 51–52
Information Officer, 79–80
Information Resources Management Service (IRMS) of GSA, 115
Innovative ideas, 39
Innovative Products Research & Services, 228
Insurance
 for business, 30–31
 financing invention with policies, 75
International Invention Register, 233
International New Product Newsletter, 232
Interview on hiring outside marketer, 134–137
Invent!, 232–233
Invention brokers, 98–99
Invention marketing flowchart, 5
Inventions
 defined, 4–6
 ideas, types of, 6–7
Inventor, defined, 3–4
Inventors Association of New England, 228
Inventors' clubs, 75
Inventors' Council, 228–229
Inventor's Desktop Companion (Levy), 99
Inventors' Digest, 226, 233
Inventor's Journal
 Idea Book and, 10
 protecting invention with, 50
Inventor's Marketing Association, 229
Inventors' shows, 41
Inventors USA, 229
Inventors Workshop International, 229–230
Inventory control, 12
Inverted pyramid style of news release, 159
Invitations
 advertising with, 163
 for bids by U.S. government, 112
 to trade shows, 220
IRMS of GSA, 115

J
Jobbers
 defined, 104
 discounts, 85
Jobs, Steve, 7
Journals. *See* Magazines
Junior college classes. *See* Community college classes

K
Karras Seminars, 18
Key man insurance, 30
Kiwanis Clubs, 15, 151
Konikow, Robert B., 224

L
Labor-saving devices, 6

Laser Sword, 6
Leadership skills, 11
Leads club, 17
Leased departments, sales to, 108–109
Lectures for publicity, 151
Lenders. *See* Financing invention
Lesko, Matthew, 76
Letterhead, 29–30
Letter of Patent/Letters Patent, 57
Letters. *See* Sales letters
Letters to the editor, 154
Levinson, Jay, 185
Levy, Dick, 99
Liability insurance, 30
Libel Manual, 14
Libraries
  ideas from, 41
  news release research in, 149
  Reader's Guide to Periodical Literature (RGPL), 149
  research skills and, 18
  Trademark Search Library, 55
Library of Congress marketing surveys, 236
Licenses and permits, 26–28
  federal licenses, 28
  fictitious business name permits, 26–27
  foreign license, ideas from, 41
  local licenses, 27
  patent licensing, 63
  special licenses, 28
  state licenses, 27–28
Licensing brokers, 41
Licensing of patent, 100–101
Life insurance, 30
Lifestyle, 196
Limited-facility wholesalers, 104
List brokers, 180

Local government financing, 76
  SBA loans and, 77
Local licenses, 27
Local magazines, 190
Local newspapers
  advertising in, 165
  for initial test, 184
  publicity and, 142
LOC programming, 168
Lorenzen, Jim, 179–181
Los Angeles Magazine, 190, 194
Los Angeles Marathon, 88
Los Angeles Times, 165
Low-Income Entrepreneurs, SBA Loans for, 77

M
Maas, Jane, 208
Magazines, 189–194
  categories of, 189–190
  characteristics of, 190–192
  contacts in, 144
  CPM (Cost Per Thousand) rates, 194
  customer coverage and, 191–192
  for inventors, 232–233
  key points for advertising in, 192–193
  market selectivity and, 190–191
  media cost-comparison grid, 202
  networks of, 194
  shelf-life of, 192
  typical rate card for, 201
Mail. *See also* Direct-mail package
  advertising through, 162–164
  trade show promotion through, 220
Mail-order advertising, 106–107, 204–208. *See also* Direct-mail package
  cost of investment, 205

  headlines in, 207
  houses, mail-order, 85
  mailing lists, development of, 205–206
  wholesalers, 104
Mail Order Business Directory, 106
Mail-Order Rule, 204
Make-ready printing plates, 29–30
Mancuso, Joseph R., 36, 80
Mancuso's Small Business Resource Guide, 80
Mandatory Source Programs of U.S. government, 114–115
Manufacturers
  catalogs, 44
  financing invention with, 75
Manufacturer's Agents National Association (MANA), 105
  marketing by, 134
Manufacturer's reps
  for distribution, 105
  marketing by, 134
Manufacturing invention, 93–95
  farming it out, 94–95
  subcontracting, 94
Marketing. *See also* Agencies for marketing; Target markets
  budget for, 85–87
  business plan for, 36
  competition and, 127–128
  competitor evaluation checklist, 131
  customer/product analysis, 130
  defined, 124
  empty niche, capitalizing on, 127, 129
  enthusiasm and, 40
  farming out marketing, 132–134
  feasibility study on, 45
  ideal products and, 40
  important three questions of, 129–131

Marketing (*continued*)
  interview on hiring outside marketer, 134–137
  key points on, 126–129
  mission statement for, 131–132
  outside help for, 132
  packaging and, 234–235
  planning for, 81–82
  positioning, 129
  scheduling of, 87–90
  skills for, 13–18
  test-marketing, 46–47
  umbrella, 125
  unique selling proposition (USP), 126–127
Marketing consultants, 132–133
Marketing umbrella, 125
Martin, Charles L., 43, 130, 313
Master distributors, 105
Mastermind groups, 17
Match the competition budgeting, 86
Maturation stage for sales, 103
Measurable goals, 22
Media. *See also* Advertising; Publicity; specific media
  cost-comparison grid, 202
  finding right media for advertising, 200–203
  myths about, 139–140
Media advisory, 152
Media analysis grid, 157, 250
Media Contact File, 143
  clip sheets in, 155–156
  directories for, 147
Media kit, 148, 158, 201
  for trade shows, 221
Media tracking sheet, 157, 251
Microcomputer Marketplace, 109–110
Military Agencies, sales to, 119

Mind map for generation of ideas, 43
Minnesota Inventors Congress, 230
Minorities
  government, sales to, 114–115
  media directories for, 145
Mission statement, 131–132
Models
  idea, scale model of, 46
  selling your invention with, 97
Monthly budget guideline chart, 88
MTV, 196

N
Names
  company name, selection of, 25–26
  fictitious business name permits, 26–27
  of invention, 239–240
National Association of Exposition Managers, 224
National Bureau of Standards, 77
National Business Incubator Association, 75
National Trade and Professional Associations Directory, 134
Navy, Department of, 118
NBS Office of Energy-Related Inventions, 79
Negotiating skills, 18
Negotiating Techniques, Economics Press, 18
Networking, 225–233
  publicity and, 151
  skills, 16–18
Newsletter Directory, 147
Newsletters, 124, 151
  advertising with, 163
  directories for, 147
  for inventors, 232–233

  self-syndicated articles in, 155
  trade show promotion through, 220
Newspaper Advertising Handbook (Watkins), 208
Newspapers, 165, 181–189. *See also* Local newspapers
  advantages of advertising in, 182–183
  choosing papers for ads, 184–185
  circulation of, 184–185
  clip sheets from, 155
  color ads, 189
  contacts in, 144
  co-op advertising, 183
  disadvantages of advertising in, 183–184
  discount rates, 187–188
  display advertising, 186–187
  flat rates, 187
  frequency discount, 188
  frequency of advertising in, 185
  general display rates, 186
  ideas from, 41, 42
  media cost-comparison grid, 202
  myths about, 139–140
  negotiating position of ads, 188
  positioning ads in, 188
  preferred position ads, 188
  readership and, 185
  reproduction quality in ads, 184
  ROP ads, 188
  sections, choice of, 188
  self-syndicated articles in, 155
  shelf-life of ads, 183
  special event scheduling and, 88
  types of ads, 185–187
  typical rate card for, 202

News releases, 148–150, 153, 158–160
   Five Ws and H formula for, 159–160
   inverted pyramid style of, 159
   project tracking sheet for, 246
   rules for, 159–160
   trade show promotion through, 220–221
Newsworthiness of invention, 140
New York, 190
New York Times, 165
Nightingale-Conant, 18
Nondisclosure agreement, 50–51
   selling your invention and, 96
Nonexclusive patent license, 101
Notice of Allowance of patent, 61

O

Oasis Press, The, 23
Objection Answering, sales and, 15
Objective-and-task method budgeting, 86–87
Office of Planning and Coordination, 79
Ogilvie, David, 129
Oklahoma Inventors Congress, 230–231
Old products, new uses for, 6
Olmi, Antonio M., 25
101 Ideas for News Releases, 158
Op-Ed pieces, 154
Orange Coast Daily Pilot, 165
Orange Coast Magazine, 194
Orange County Register, 165
Organizational outline, 36
Organizational skills, 16
   checklist for, 11
Organizations. *See* Associations and clubs

Organized Executive, The (Winston), 16
Originality of invention, 61
Outdoor signs, 169
Overseas selling
   Information Officer for, 79–80
   marketing for, 126
Oxbridge Directory of Newsletters, 147

P

Package plan for radio advertising, 198
Packaging invention, 234–240
   color, use of, 237
   costs of, 85
   displays, types of, 238
   marketing plan and, 234–235
   naming your invention, 239–240
   rules for, 236
   shape of packaging, 237–238
   size of packaging, 238
   for target markets, 235–236
   types of packaging, 239
   visual messages in, 236–238
Parties for invention, 163
Partnerships, 24
   key man insurance for, 30
Patent agents, 60
Patent and Trademark Office (PTO), 52, 57
   definition of invention by, 4
   guidelines for patents, 60–61
   patent attorneys, list of, 50
   writing skills and, 13
Patent attorneys
   Patent and Trademark Office (PTO) list of, 50

   registered patent attorneys, 60
   use of, 57
Patent licensing, 100–101
Patents
   action notice on, 61–62
   agency granting, 52
   amendments of, 62
   assignment of, 63
   defined, 57
   definition of invention for, 4
   developing ideas from, 41
   diagram of patent process, 58
   duration of, 57, 59
   fees for, 62
   guidelines of PTO for, 60–61
   Letter of/Letters, 57
   licensing of, 63, 100–101
   Notice of Allowance of patent, 61
   parts of application for, 57
   procedure for filing, 59, 60
   rejection of application, 59
   sale of patent, 100–101
   selling your invention, patent application and, 97
   trademark search and application process compared, 62
Patent search, 59, 62
   trademark search compared, 62
Payroll, 12
PBS of GSA, 115
Pennysaver, 190
People skills, 11
Percentage of sales budgeting, 86
Personal financial statement, 31–32
Personal funds, financing with, 75–76
Phelon's Directory, 107–108

Photograph of invention with news release, 150
for publicity, 155
Planning. *See also* Business plan
reasons for, 81–82
Plastic bags, packaging in, 239
Plateau stage for sales, 103
PLP (Preferred Lenders Program (SBA)), 73
PMS colors. *See* Colors
Point-of-purchase (POP) signs, 169
Polks Bank Directory, 73
Positioning, 129
newspaper ads, 188
Positioning: The Battle for Your Mind (Ries & Trout), 81–82
Practicality of invention, 6, 61
Pratt, Stanley, 78
Preempted spot, 197
Preferential procurement programs of U.S. government, 113–115
Preferred Lenders Program (PLP) (SBA), 73
Premiums, 109
Press. *See* Media
Press conferences, 151–152
Press kit. *See* Media kit
Prime Ticket, 196
cost of advertising on, 197
Printing costs, 29–30
Private patents, 41
Private stock placement, financing with, 76
Problem-solving ideas, 7
Proctor and Gamble, 40
Production plan, 36
Product knowledge, sales and, 15
Product liability insurance, 30
Products. *See also* Ideas
evaluation of, 43
ideal products, 40

Profitability
break-even analysis and, 84
feasibility study on, 45
Program brokering on radio, 167
Projected statement of sales and expenses for one year, 33
Promotion schedule, sample calendar for, 89
Protecting invention, 48–66. *See also* Copyrights; Patents; Trademarks
attorneys for, 49–50
Disclosure Document Program for patent, 59
informal protection, 51–52
keys to, 50–51
restrictions on, 49
self-registration of copyright, 52
selling your invention and, 96
small business consultants for, 49–50
types of protection, 51–52
Prototypes
development of, 45–46
patents of, 61
selling your invention with, 97
Public Buildings Service (PBS) of GSA, 115
Public domain, patents in, 59
Publicity, 138–160. *See also* Advertising; Media Contact File; News releases
directories for media, 145–147
establishing media relationships, 141–144
local press, 142
in magazines, 190
Media Analysis Grid, 157, 250

myths about media, 139–140
news releases, 148–150
project tracking sheet, form of, 243–250
Publicity Project Tracking Sheet, 145, 243–250
researching the media, 145
resident expert for media, 142
right person, contacting the, 144
salability of story, 140–141
samples for, 148
selection of media resources, 148
sending news release, 149–150
strategies for, 150–152
tools, 152–157
tracking results of, 150, 243–250
for trade shows, 220–222
Publicity Club of Los Angeles Directory, 151
Publicity Project Tracking Sheet, 145, 243–250
Public relations agencies, marketing by, 133–134
Public service announcements, 153
project tracking sheet for, 247

Q
Question-and-answer formats, 155–156

R
Rack jobbers, 104
Radio, 166–167. *See also* Publicity
advantages and disadvantages of advertising on, 194–195
commercials/spot buys, 167

Radio (*continued*)
  contacts in, 144
  directories, 147
  ideas from, 41
  key questions about advertising on, 195–196
  length of commercial on, 198
  media cost-comparison grid, 202
  program brokering, 167
  public service announcements on, 153
  rate cards for advertising on, 196
  segments, advertising on, 167
  sponsorships, 166, 203
  spot announcements, 197–198
  success keys for advertising on, 198–199
  tactics for buying spots, 200
  talk shows, advertising on, 167
  time segments for ads, 199
  typical rate card for, 203
Reader's Digest, 175
Reader's Guide to Periodical Literature (RGPL), 149
Real estate equity loan financing, 76
Reasons for inventing, 7–8
Regional magazines, 190
Register of Copyrights, 52, 66
Reinventing uses for old products, 6
Relationship Selling: How to Get and Keep Customers (Cathcart), 16
Relatives. *See* Family
Repeat sales potential, 40
  feasibility study on, 45
  GSA bid list and, 113

Request for proposal (RFP) by U.S. government, 112
Resale numbers, 27–28
Research skills, 18
Resident buyers, sales to, 108–109
Responsibility skills, 11
Retail displays
  advertising, 186
  packaging and types of, 238
Retailers, 102–103
  Fairchild's Financial Manual of Retailers, 108
Ries, Al, 81–82
Right Writer, 175
Role-playing for negotiating, 18
Roman, Kenneth, 208
ROP ads in newspapers, 188
ROS (run of schedule) spot, 198
Rotary, 15, 151

S
Sales journal, 83
Sales letters, 163
  in direct-mail package, 171, 174–175
Sales price, setting of, 84–85
Sales reps. *See* Manufacturer's reps
Sales taxes, quarterly payment of, 28
Sample runs for prototype, 47
Samples, advertising with, 164
San Diego Business, 190
San Diego Magazine, 194
Savings account collateral, 72
Savings and loan associations, financing with, 72–73
SBA. *See* Small Business Administration (SBA)

SBA Answerdesk, 73
Scale models. *See* Models
Scheduling, 87–90
  promotion schedule, sample calendar for, 89
  for public relations, 150
Seasonal events
  and marketing, 127
  scheduling and, 88
Secrets of Power Negotiating, The (Dawson), 18
Selecting the Legal Structure of Your Firm (Olmi), 25
Self-mailers, 163
  in direct-mail package, 176
Self-registration of copyright, 52
Self-starters, 11
Self-syndicated articles, 155
Selling out, 93–101
Selling skills, 15–16
Selling your invention, 95–96. *See also* Distribution; Publicity; United States government
  disclosure form for, 95–96
  invention brokers, 98–99
  patents, sale of, 100–101
  prospective buyers, finding of, 96–99
  sources of help for, 99–100
Seminars
  advertising at, 169
  for negotiating skills, 18
  publicity through, 151
  on writing, 14
Service marks distinguished from trademarks, 53
Sheetmetal specialists, prototypes by, 46
Shelf displays, 238

Short-term goals, 21
Shrink-wrap packaging, 239
SIC (Standard Industrial Classification), 44
Signs, advertising by, 169
Skills for success, 9–18
Skywriting, 169
Small and Disadvantaged Business Utilization Office, 114
Small Business Administration (SBA), 10
  bookkeeping skills, 13
  financing through, 73, 77
  government sales, help with, 117
  Incubator Times, 75
  insurance checklist, 30–31
  SBA Answerdesk, 73
  small business innovation research programs, 77
Small business consultants, 49–50
Small business innovation research (SBIR) programs
  government sales, help with, 117
  SBA loans, 77
  USDA loans, 77
Small business investment companies (SBIC)
  financing through, 77
  ideas from, 41
Small Business Set-Asides, 114
Socially and Economically Disadvantaged Businesses, 114
Society of Minnesota Inventors, 231
Sole proprietorships, 23
Solicitation mailing list application, form of, 116
Speaking skills, 14–15
Special events, scheduling and, 88

Special-interest newspapers, 184
Specialties, 109
  advertising specialties, 169
Specialty directories, advertising in, 165–166
Speeches for publicity, 151
Sponsorships in radio or television, 166, 203
Spontaneous ideas, 7
Sports Channel, 196
Sports Marketplace, 109–110
Spot announcements, 197–198
Standard and Poor's Register of Corporations, 95
Standard Rate and Data Service (SRDS), 44, 146, 165
  broadcast media advertising reference, 166
  list brokers, selection of, 180
  magazine volumes, 190
  media, choice of, 201
Starting and Operating a Business series, 23
Starting Your New Business (Martin), 43, 130, 131
Star Wars, 6
State licenses, 27–28
State programs, financing through, 76
Stationery, 29–30
Stock placement, financing with, 76
Stores. *See also* Retail displays; Retailers
  directories of, 107–108
  ideas from, 41
  newness of idea, checking for, 44
  prototypes at, 47
Strunke, William, Jr., 14
Subchapter S corporations, 25
Subcontracting, 94

Subcontracts—small and Small Disadvantaged Businesses, 114
Suburban newspapers. *See* Local newspapers
Success, 190
Success skills, 9–18
Surety bonds for resale numbers, 28
Symbols for trademarks, 55

T
Talk shows, 151
  question-and-answer formats for, 155–156
  radio talk shows, advertising on, 167
Targeted goals, 22
Target markets
  budgeting and, 87
  cable television for, 196
  finding your, 124–126
  packaging invention for, 235–236
Taxes
  business-types and, 23–25
  sales taxes, quarterly payment of, 28
Technical writing skills, 13–14
Technology Transfer Society, 231
Telemarketing, 170
Television, 167–168. *See also* Cable television; Publicity
  advantages and disadvantages of advertising on, 195
  contacts in, 144
  directories, 147
  ideas from, 41
  key questions about advertising on, 195–196
  length of commercial on, 198
  LOC programming on cable TV, 168

Index 271

Television (*continued*)
  media cost-comparison grid, 202
  public service announcements on, 153
  rate cards for advertising on, 196
  sponsorships, 166, 203
  spot announcements, 197-198
  tactics for buying spots, 200
  time segments for ads, 200
  tips for good commercials, 199
Test-marketing, 46-47
Thomas Register of Manufacturers, 44, 94, 95
Three-color printing, 29
Time management skills, 16
Toastmasters International, 14-15
Trade directories, 95
Trade magazines
  advertising in, 189
  ADWEEK, 134
  ideas from, 41
  newness of idea, checking for, 44
  self-syndicated articles in, 155
  selling your invention through, 95
Trademark Official Gazette, 55
Trademarks
  agency granting, 52
  checking on, 28-29
  costs of, 55, 57
  defined, 52
  diagram of trademark process, 56
  duration of, 53
  filing registration, method for, 53-55
  form of application for, 54
  patent process compared, 62
  service marks distinguished, 53
  symbols for, 55
Trademark search, 29, 55
  patent search compared, 62
Trademark Search Library, 55
Trademark Trial and Appeal Board, 55
Trade Show Bureau, 224
Trade shows, 109, 209-224
  advertising at, 170
  at-show promotion, 221
  boothmanship at, 224
  budget of show, 212-214
  checklist for logistics, 223
  checklist for successful booth, 217-218
  design of show, 212
  display suggestions for, 218-219
  exhibit booth at, 214-218
  ideas from, 41
  location of booth, 214-215
  logistics of, 222-223
  planning the show, 211
  post-show promotion, 221-222
  pre-show meetings, 223
  press kit at, 158
  prototype display at, 47
  publicity for, 220-222
  resources for, 224
  sales literature, display of, 219
  schedule of show, 211
  setting objectives for, 210
  size of booth, 216-217
  staffing needs for, 215-216
  theme of show, 211-212
  union labor at, 222
Tradeshow Week Data Book, 211
Tradeshow Week Magazine, 211, 224
Transportation to trade shows, 222
Trout, Jack, 81-82
Turner, Ted, 196
TV. *See* Television
Two-color printing, 29

U
UHF television, 196
Union labor at trade shows, 222
Unique selling proposition (USP), 126-127
United States government. *See also* Bidders' lists; Small Business Administration (SBA)
  bids, advertising for, 112
  buyers, location of, 113-115
  contracts, types of, 112
  financing through, 76-77
  fixed-price contracts with, 112
  licenses required by, 28
  negotiation, buying by, 112
  patents owned by, 41
  preferential procurement programs, 113-115
  purchasing methods of, 111-112
  qualified products lists, 112
  request for proposal (RFP), 112
  solicitation mailing list application, form of, 116
United States Government Purchasing and Sales Directory, 113, 117
Unknown inventions, 61
Usefulness of inventions, 61

V
Valdez, Gene, 78
Venture capital clubs, financing through, 78

Venture capital companies, 78
Venture Capital Network, 78
VHF television, 196
Vietnam Veterans, U.S. government contracts to, 114
Von Clausewitz, Carl, 81–82
Voucher register, 83

W
Wall displays, 238
Wants and needs, 39–40
Watkins, Don, 208
Weekly newspapers, advertising in, 165
White, E. B., 14
Wholesalers, 102–103
    defined, 103
    discounts, 85
    distributors, 104–105
    full-facility wholesalers, 103–104
    limited-facility wholesalers, 104
Window posters, 169
Winston, Stephanie, 16
WISC (Wisconsin Innovation Service Center), 99, 232
Woman Magazine, 191
Women-Owned Businesses, 114
Workability of invention, 61
Working Mother Magazine, 191
Working Press of the Nation (WPN), 146, 147
Working Woman Magazine, 191
Worksheet for invention ideas, 7
Workshops, publicity through, 151
Work skills, 11
World Series, 88
Wozniak, Steve, 7
Writer Magazine, The, 14
Writer's Digest, 14
Writer's Market, 146
    for magazines, 190–191
Writer's Yellow Pages, The, 14
    grant proposal writers, 74
Writing
    skills, 13–14
    software for, 175

X
Xerox Foundation, 79

Y
Yellow pages, advertising in, 170
You Can Sell Anything by Telephone (Goodman), 170

Z
Ziglar, Zig, 16

Seminars Offered by Reece A. Franklin

INVENTOR'S MARKETING WORKSHOP

An intensive one-day course based on the book *How to Sell and Promote Your Idea, Project, or Invention.* Includes sessions on idea generation, prototypes, patents, trademarks, copyrights, financing and budgeting, selling out for big profits, selling to the U.S. government, marketing, public relations, advertising, trade show marketing, networking—and more! Now that you've bought the book, take the seminar for some hands-on instruction tailored to *your* invention!

HOW TO GET FREE PUBLICITY

Number One on the hit parade of seminar attendees the past 5 years in a row! Greatly expanded and revised, with supplemental information from local media sources. Perfect for small businesses and nonprofits. Half day or full day. Topics include news release preparation, elements of a press kit, how to book radio and television interviews, making editors want your information, dealing effectively with the media—and more!

HOW TO ADVERTISE AND PROMOTE YOUR SMALL BUSINESS

This one-day, intensive seminar is designed for those small businesses who realize the need for *effective* advertising and promotion plans but don't know where to start. Topics include the elements of a good ad, how to set up an ad budget, how to get *more* from your ad space, which media to use and why, cost-effective promotions that work, how to get *free* layout design and artwork—and more!

DESIGNING EFFECTIVE SMALL BUSINESS ADS

This course, offered in response to student requests, gives you everything you need to design ads that work. This half-day, intensive workshop includes producing eye-catching ads that sell, the basic tools for success, how to write good ad copy, how to write headlines that make money, how to upgrade your company image using small space effectively—and more!

For complete information on these seminars or on how I can help you with your idea, project, or invention, contact me:

REECE FRANKLIN AND ASSOCIATES
P.O. Box 2667
Chino, CA 91708-2667
(909) 393-8525

ORDER FORM

Reece Franklin & Associates
P.O. Box 2667
Chino, CA 91708-2667
(714) 393-8525

Please send me copies of the following publications:

_____ copies of *How to Sell and Promote...* @ $16.95 each

_____ copies of *101 Ideas for News Releases* @ $4.95

_____ copies of *Promo Ideas for Retail Stores* @ $6.95

_____ copies of *How to Market Your Home Based Business* @ $17.95

Please send me copies of the following audio tapes:

_____ copies of *Marketing for Inventors* @ $9.95

_____ copies of *Publicity for Inventors* @ $9.95

I understand if I'm not satisfied, I may return any book in resellable condition for a full refund.

_____ I am interested in attending your **INVENTOR'S MARKETING WORKSHOPS**. Please send me details.

_____ I am interested in attending your other seminars. Please send me a brochure.

Name: _____

Address: _____

City: _____

State: _____ Zip: _____

Californians: Please add 7.75% sales tax.

Shipping: $2.50 for the first book and $1.00 for each additional book.

_____ I can't wait 3–4 weeks for book rate. Here is $3.00 per book for Air Mail.